PRACTICAL COURSE IN PHYSICS

For First Year B.Sc. PHY-113 & 123 : Semester I & II
As per New Syllabus CBCS Pattern Credit – 2, June 2019

Dr. P. P. BARDAPURKAR
M.Sc., Ph.D.
Assistant Professor,
Department of Physics,
S.N. Arts, D.J. Malpani Commerce & B.N. Sarda
Science College, Sangamner, Ahmednagar.

Dr. N. P. BARDE
M.Sc., SET, Ph.D.
Assistant Professor,
Department of Physics,
Badrinarayan Barwale
Mahavidyalaya, Jalna.

Dr. S. A. AROTE
M.Sc., SET, Ph.D.
Assistant Professor,
Department of Physics,
S.N. Arts, D.J. Malpani Commerce &
B.N. Sarda Science College,
Sangamner, Ahmednagar.

Dr. D. L. GAPALE
M.Sc., M.Phil., Ph.D
Assistant Professor,
Department of Physics,
S.N. Arts, D.J. Malpani Commerce &
B.N. Sarda Science College,
Sangamner, Ahmednagar.

Dr. S. N. DALVI
M.Sc., Ph.D.
Assistant Professor & Head
Department of Physics,
S.N. Arts, D.J. Malpani Commerce &
B.N. Sarda Science College,
Sangamner, Ahmednagar.

Dr. B. M. Palve
M.Sc., SET, Ph.D.
Assistant Professor,
Department of Physics,
S.N. Arts, D.J. Malpani Commerce &
B.N. Sarda Science College,
Sangamner, Ahmednagar.

N5004

F.Y.B.Sc. Practical Course in Physics   ISBN 978-93-89406-98-6

First Edition : **August 2019**

© : **Dr. P. P. Bardapurkar**

Published By :

NIRALI PRAKASHAN

Abhyudaya Pragati, 1312 Shivaji Nagar,
Off J.M. Road, PUNE - 411005
Tel - (020) 25512336/37/39. Fax - 25511379
Email : niralipune@pragationline.com

➢ DISTRIBUTION CENTRES

PUNE

Nirali Prakashan : 119, Budhwar Peth, Jogeshwari Mandir Lane, Pune 411002, Maharashtra
Tel : (020) 2445 2044, 66022708, Fax : (020) 2445 1538
Email : bookorder@pragationline.com, niralilocal@pragationline.com

Nirali Prakashan : S. No. 28/27, Dhyari, Near Pari Company, Pune 411041
Tel : (020) 24690204 Fax : (020) 24690316
Email : dhyari@pragationline.com, bookorder@pragationline.com

MUMBAI

Nirali Prakashan : 385, S.V.P. Road, Rasdhara Co-op. Hsg. Society Ltd.,
Girgaum, Mumbai 400004, Maharashtra
Tel : (022) 2385 6339 / 2386 9976, Fax : (022) 2386 9976
Email : niralimumbai@pragationline.com

➢ DISTRIBUTION BRANCHES

JALGAON

Nirali Prakashan : 34, V. V. Golani Market, Navi Peth, Jalgaon 425001,
Maharashtra, Tel : (0257) 222 0395, Mob : 94234 91860

KOLHAPUR

Nirali Prakashan : New Mahadvar Road, Kedar Plaza, 1st Floor Opp. IDBI Bank
Kolhapur 416 012, Maharashtra. Mob : 9850046155

NAGPUR

Pratibha Book Distributors : Above Maratha Mandir, Shop No. 3, First Floor,
Rani Jhanshi Square, Sitabuldi, Nagpur 440012, Maharashtra
Tel : (0712) 254 7129

DELHI

Nirali Prakashan : 4593/21, Basement, Aggarwal Lane 15, Ansari Road, Daryaganj
Near Times of India Building, New Delhi 110002 Mob : 08505972553

BENGALURU

Nirali Prakashan : Maitri, Ground Floor, Java Apartments No. 99
6th cross, 6th main Malleswaram,
Benguluru – 560003   Mob : 9449043034
Email: niralibangalore@pragationline.com

CHENNAI

Pragati Books : 9/1, Montieth Road, Behind Taas Mahal, Egmore,
Chennai 600008 Tamil Nadu, Tel : (044) 6518 3535,
Mob : 94440 01782 / 98450 21552 / 98805 82331,
Email : bharatsavla@yahoo.com

niralipune@pragationline.com | www.pragationline.com

Also find us on 🅕 www.facebook.com/niralibooks

PREFACE

This book is written according to the revised choice based credit system syllabus in **Practical Course in Physics** for *B.Sc. First year* course as suggested by Savitribai Phule Pune University. It will serve as an excellent introduction at undergraduate level.

We are happy to present this book covering the need of a student during the experimental work at the laboratory. The standard of the book is maintained keeping the level of First Year B.Sc. course in terms of the steps required for performing the experiments. However the format of procedures to perform the experiment, observation tables, theory, viva-voce questionnaires etc. are provided wherever it is necessary for deep understanding.

Utmost care has been taken to explain the steps for performing practicals with illustrative figures and circuit diagrams used where absolutely required.

We are grateful to Dr. Sanjay Malpani; Chairman and all the office bearers; S. P. Sanstha, Sangamner for their consistent encouragement.

We are also thankful to Principal Dr. K. K. Deshmukh and Vice Principals, Sangamner College, Dr. P. M. Kokne, Badrinarayan Barwale College, Jalna for their kind support.

Our sincere thanks are to Shri. Dineshbhai Furia, Shri. Jignesh Furia, Shri. M. P. Munde and whole staff of Nirali Prakashan, especially Mr. Nilesh Deshmukh, Mr. Kiran Velankar, Mr. Ravi Walodare, Mrs. Yojana G. Deshpande and Ms. Chaitali Takle for making this effort possible.

We also thankful to all the teachers and students who have extended their enormous help to get the work in the present form, for their unending support and valuable ideas!

We hope that this book will be useful to the students and teaching staff at the graduation level. Suggestions for improvement are most welcome........

– **Authors**

SYLLABUS

Course code and title: PHY-113 Physics Laboratory 1A
Practical: 08 (Credits-1.5)

Section I- Mechanics and Properties of Matter

(1) Study and use of various measuring Instruments: (a) Vernier caliper (b) Micrometer Screw Gauge (c) Travelling Microscope

(2) Study of Modulus of Rigidity of wire using Torsional Oscillations

(3) Determination of coefficient of Viscosity by Poiseuille's method

(4) Determination of "Y" and "η" by flat spiral spring

(5) Determination of "Y" by bending method.

(6) Study of surface tension by Jaeger's method

(7) Study of Poisson's ratio of rubber using rubber tube /rubber chord

(8) Study of surface tension of liquid using Fergusson Method

Section II-Physics Principles and Applications

(1) Study of Spectrometer and determination of angle of prism

(2) Study of Spectrometer calibration and determination of refractive indices of different colors

(3) Study of divergence of LASER beam

(4) Study of total internal reflection using LASER

(5) Determination of Plank's constant

(6) Determination of wavelength of LASER light by plane diffraction grating

(7) Study of I-V characteristics of solar cell

> **Note :** *Any four experiments from each section be conducted during the semester*

Course code and title: PHY-123 Physics Laboratory 1B
Practical: 08 (Credits-1.5)

Section I- Heat and Thermodynamics

(1) Interpretation of Isothermal and Adiabatic curve on P-V diagram and theoretical study of Carnot's cycle by drawing graphs of Isothermal and Adiabatic curves

(2) Study of temperature coefficient of Thermistor

(3) Study of Thermocouple and determination of inversion temperature

(4)   Study of thermal conductivity by Lee's method

(5)   Study of specific heat of Graphite

(6)   Study of Solar constant

(7)   Determination of calorific values of different fuels

Section II- Electricity and Magnetism

(1)   Study of charging and discharging of capacitor

(2)   Study of LR circuit

(3)   Study of LCR circuit

(4)   Study of Kirchhoff's laws

(5)   Diode characteristics.

(6)   Study of Voltmeter, Ammeter and Multimeter (AC, DC, ranges and least count)

(7)   Determination of frequency of AC mains

(8)   Comparison of capacitor using DeSauty's method

Note : *Any four experiments from each section be conducted during the semester.*

CONTENTS

Course code and title: PHY-113 Physics Laboratory 1A
Practical: 08 (Credits-1.5)

Course code and title: PHY-123 Physics Laboratory 1B
Practical: 08 (Credits-1.5)

Expt. No.	Title	Date	Page No.	Teacher's Signature
Section I- Heat and Thermodynamics			**78 - 111**	
1.	Interpretation of Isothermal and Adiabatic curve on P-V diagram and theoretical study of Carnot's cycle by drawing graphs of Isothermal and Adiabatic curves		78	
2.	Study of temperature coefficient of Thermistor		83	
3.	Study of Thermocouple and determination of inversion temperature		88	
4.	Study of thermal conductivity by Lee's method		93	
5.	Study of specific heat of Graphite		98	
6.	Study of Solar constant		104	
7.	Determination of calorific values of different fuels		108	
Section II-Electricity and Magnetism			**112 -152**	
1.	Study of charging and discharging of a capacitor		112	
2.	Study of LR circuit		117	
3.	Study of LCR circuit		122	
4.	Study of Kirchhoff's laws		128	
5.	Diode characteristics		133	
6.	Study of Voltmeter, Ammeter and Multimeter (AC/DC ranges and least count)		138	
7.	Determination of frequency of AC mains		144	
8.	Comparison of capacitors using Desauty's method		149	

❖ ❖ ❖

I – 1

/ /

Experiment performed on

STUDY AND USE OF VARIOUS MEASURING INSTRUMENTS

 **Aim :** To study range and least count of instruments, measurements using various instruments and error analysis.

 **Apparatus :** Meter scale, Vernier calliper, Micrometer screw gauge, Travelling microscope, Cylindrical pipe, Wire, Glass tube, Magnifying glass, etc.

Figure :

Vernier Calliper   **Micrometer S. G.**   **Travelling microscope**

Fig. 1.1

 **Procedure :**

1. Find out the range of the given instrument.
2. Note down the smallest division on main scale and total number of divisions on the vernier scale. With this, calculate the least count of the given instrument.
3. Record the readings in the tabular form.
4. Repeat the above procedure for all the given measuring instruments.
5. Report the results in a given format.

Observations and Observation Table :

1. Vernier Calliper :

- Range: From ………….. to ………….. cm
- Smallest division on main scale = …………………….. cm

Total number of divisions on the vernier scale = ……………………..

$$\text{L.C. of vernier calliper} = \frac{\text{Smallest division on main scale}}{\text{Total number of divisions on vernier scale}} = \text{………….. cm}$$

Measurements using Vernier Calliper:

To determine outer diameter of a cylindrical pipe				
Obs. No.	Main scale reading (a) (cm)	Vernier scale reading (b) (div)	Total reading T.R. = a + (b × L.C.) (cm)	Mean (cm)
1.				
2.				
3.				

To determine inner diameter of a cylindrical pipe				
Obs. No.	Main scale reading (a) (cm)	Vernier scale reading (b) (div)	Total reading T.R. = a + (b × L.C.) (cm)	Mean (cm)
1.				
2.				
3.				

2. Micrometer Screw Gauge:

- Range: From …………. to ………….. cm
- Smallest division on main scale = …………………….. cm
- Total number of rotations to travel smallest division on main scale = ……………

Pitch of the screw =

$$\frac{\text{Value of smallest division on main scale}}{\text{Total number of rotations to travel smallest division on main scale}} = \text{…………. cm}$$

$$\text{L.C. of screw gauge} = \frac{\text{Pitch of the screw}}{\text{Total number of divisions on circular scale}} = \text{………….. cm}$$

3. Travelling Microscope:

- Range: From …………. to ………….. cm
- Smallest division on main scale = …………………….. cm
- Total number of divisions on the vernier scale = ……………………..

$$\text{L.C. of travelling microscope} = \frac{\text{Value of smallest division on main scale.}}{\text{Total number of divisions on vernier scale}}$$

$$= \text{………….. cm}$$

Measurements using Travelling Microscope:

		To determine inner diameter of a glass tube				
Obs. No.	T.M. focused at	Main scale reading (a) (cm)	Vernier scale reading (b) (div)	Total reading T.R. = a + (b × L.C.) (cm)	Mean (cm)	Diameter (D) (cm) $D = \left\| d_1 - d_2 \right\|$
1.	1st edge				$d_1 =$	
2.						
1.	2nd edge				$d_2 =$	
2.						

 **Result :**

Instruments		Results
Vernier Calliper	Range	= From ....to .... cm
	Least count	= .... cm
	Outer diameter of the pipe	= .... cm
	Inner diameter of the pipe	= .... cm
Micrometer Screw Gauge	Range	= From ....to .... cm
	Least count	= .... cm
	Diameter of the wire	= .... cm
Travelling Microscope	Range	= From ....to .... cm
	Least count	= .... cm
	Inner diameter of the glass tube	= .... cm

 **Theory :**

- Experiments in Physics involve various measurements of quantities and it is most important to have these measurements as accurate and reproducible as possible. Certain basic standards of units and measurements have been established internationally.

- These are the units of length, mass and time. All the measurements in physics are related to these three fundamental units. All other units of measurements can be derived from these three fundamental units.

- In the Metric System there are two commonly used systems of measurement, one based on the Meter, Kilogram and Second (MKS) and the other on the Centimeter, Gram and Second (CGS).

- **The Meter Stick:** A meter stick, by definition, is 1 meter (m) long. Its scale is divided, and numbered, into 100 centimeters (cm). Each centimeter, in turn, is divided into 10 millimeters i.e. 1 cm = 10 mm = 10^{-2} m, 1 mm = 10^{-3} m.

- **Errors:** In general, if any two or more measurements are made of the same physical quantity, they have slight difference in their values. The difference between a measurement and the true value of the physical quantity is known as error in the measurement.

- The term "error" is different from English word "mistake." Errors are inherent in measurements and cannot be avoided. Mistakes can, and should, be avoided.

- **Random errors:** Occur in all measurements. They arise when observers approximate the last figure of the reading. These are called random, because they cannot be predicted. To minimize random error is to take the average of many readings.

- **Systematic errors:** Such mistakes are not random, but constant. Systematic errors may be due to defective equipment - for instance, an incorrectly marked ruler; or they may be due to environmental factors - for instance, the weather conditions on a particular day.

Prefix	Symbol	multiple
Deci	d	10^{-1}
Centi	c	10^{-2}
Milli	m	10^{-3}
Micro	μ	10^{-6}
Nano	n	10^{-9}
Pico	p	10^{-12}
Femto	f	10^{-15}
Atto	a	10^{-18}
Deca	da	10^{1}
Hecto	h	10^{2}
Kilo	k	10^{3}
Mega	M	10^{6}
Giga	G	10^{9}
Tera	T	10^{12}
Peta	P	10^{15}
Exa	E	10^{18}

$$\text{Percentage error} = \frac{\text{Standard value} - \text{Observed value}}{\text{Standard value}} \times 100$$

$$\text{Percentage error} = \ldots\ldots\ldots\ldots \%$$

 **Precautions :**

1. Measure the least count carefully.
2. Write appropriate units.
3. Take a proper care with significant numbers during calculations.
4. Use magnifying lens if required.

 **Check Your Grasp :**

1. *What is least count?*
2. *What is a scale?*
3. *What is meant by smallest division and range of an apparatus?*
4. *What are different measuring instruments?*
5. *State at least two applications for each of the above measuring instruments.*

 **Calculations :**

STUDY OF MODULUS OF RIGIDITY OF WIRE USING TORSIONAL OSCILLATIONS

Aim : To determine modulus of rigidity of material of a wire using a torsional pendulum.

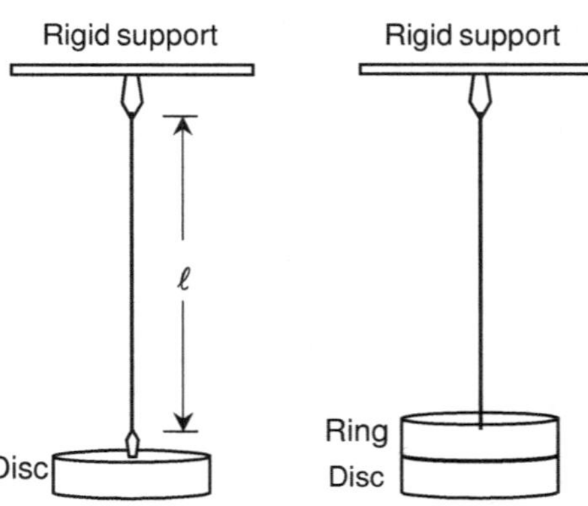

Apparatus : Disc suspended by wire, ring, stop watch, vernier calliper, micrometer screw gauge, pin, etc.

Figure :

Fig. 2.1 : Disc only   Fig. 2.2 : Disc + Ring

Procedure :

1. Suspend the wire and disc as shown in Fig. 2.1.
2. Measure the inner and outer radius of the ring. Also measure the radius of the suspended wire.
3. Give the torsional oscillation (**twist by 45°**) and measure the time (t_1) for 10 oscillations. Repeat these two times to get t_2 and t_3 for 'disc only' system. Calculate the mean time (t).
4. Calculate periodic time T_o for the **disc** ($T_o = t/10$ sec).

5. Keep the ring on the disc as shown in Fig. 2.2 and repeat the above procedure.
6. Note down the time: t_1, t_2, t_3 and find out the mean time t for 'disc + ring' system.
7. Calculate the periodic time T_1 for **ring + disc** system ($T_1 = t/10$ sec).
8. Using the obtained data, calculate the moment of inertia of the ring and modulus of rigidity of material of the wire.

 **Observations :**

- Least count of vernier calliper, L.C. = ...cm
- Least count of micrometer screw gauge, L.C. = ...cm
- Inner diameter of the ring (**d**)

  (i) d_1 = ... cm (ii) d_2 = ... cm (iii) d_3 = ... cm

  Mean diameter = d = $\dfrac{d_1 + d_2 + d_3}{3}$ .... cm

  Inner radius of the ring, r = $\dfrac{d}{2}$ = ... cm

- Outer diameter of the ring (**D**):

  (i) .... cm (ii) .... cm (iii) .... cm

  Mean diameter D = ... cm

  Outer radius of the ring R = $\dfrac{D}{2}$ = ... cm

- Mass of the ring M = ... g.
- Length of the wire l = ... cm
- Diameter of the wire (A)

  (i) A_1 = ... cm (ii) A_2 = ... cm (iii) A_3 = ... cm

  Mean diameter = A = $\dfrac{A_1 + A_2 + A_3}{3}$ .... cm

  Radius of the wire, a = $\dfrac{A}{2}$ = ... cm

 **Observation Table :**

	Suspended system	Time for 10 oscillations (t) (s)				Periodic time (T) $T = t/10$ (s)	T^2 (s²)	Difference $T_1^2 - T_0^2$ (s²)
		t_1	t_2	t_3	Mean (t)			
1	Disc only					$T_0 =$	$T_0^2 =$	
2	Disc + Ring					$T_1 =$	$T_1^2 =$	

Formulae :

(a)   M.I. of the ring, $I_r = \dfrac{M\,(R^2 + r^2)}{2}$ g cm²

(b)   M.I. of the disc, $I_d = \dfrac{I_r\,T_0^2}{(T_1^2 - T_0^2)}$ g cm²

(c)   Modulus of rigidity, $\eta = \dfrac{8\pi\ell I}{a^4\,(T_1^2 - T_0^2)}$ dyne/cm²

Result :

- M.I. of the ring, I_r = ... g cm²
- M.I. of the disc, I_d = ... g cm²
- Modulus of rigidity, η = ... dyne/cm²

Theory :

- A torsional pendulum consists of a disc-like mass suspended from a thin wire. When the mass is twisted about the axis of the wire, the wire exerts a torque on the mass, tending to rotate it back to its original position. Such a back and forth oscillation executes simple harmonic motion. This gives us an idea of moment of inertia. We try to calculate the moment of inertia of a ring.

- Consider a disc suspended from a thin wire attached to its centre. A wire is essentially inextensible, but is free to twist about its axis. When we apply force, the wire twists and causes the disc attached to it to rotate. Let θ be the angle of rotation of the disc, and let $\theta = 0°$ correspond to the case in which the wire is untwisted i.e. the disc is stationary.

- Any twisting of the wire is certainly associated with mechanical deformation. The wire resists such deformation by developing a restoring torque 'τ' which acts to restore the wire to its original untwisted state. For a given twist or deformation, the magnitude of this torque is directly proportional to the twist angle, θ.

- Hence, we can write, $\tau = -\,k\theta$

  where $k > 0$ is the torque constant of the wire. The above equation is essentially a torsional equivalent to Hooke's law. The negative sign is due to restoring force.

Precautions :

1. Ensure that the measuring instruments are free from zero error.
2. Measure diameter of the wire very carefully.
3. Don't disturb the apparatus throughout the experiment.
4. Amplitude of torsional oscillations should be small.
5. Disc of the pendulum should oscillate only in the horizontal plane.

Check Your Grasp :

1. *What are the various types of pendula?*
2. *How a torsional pendulum differs from a simple pendulum?*
3. *Define mass, radius, inertia, moment of inertia and modulus of rigidity.*
4. *What is radius of gyration?*
5. *What are stress, strain, Hook's law and twisting couple?*
6. *What do you mean by Torsional oscillations?*
7. *Define: Period of oscillation, Restoring force*
8. *Why amplitude of torsional oscillations should be small?*
9. *How will you calculate angle of shear produced in the wire?*
10. *How oscillations are established in a torsional pendulum?*

Calculations :

COEFFICIENT OF VISCOSITY BY POISEUILLE'S METHOD

Aim : To determine coefficient of viscosity of water by Poisseuille's method.

Apparatus : Long capillary tube, Constant pressure apparatus, Stop watch, Measuring cylinder, Rubber tubes, Pinch cock, Meter scale, Manometer etc.

Figure :

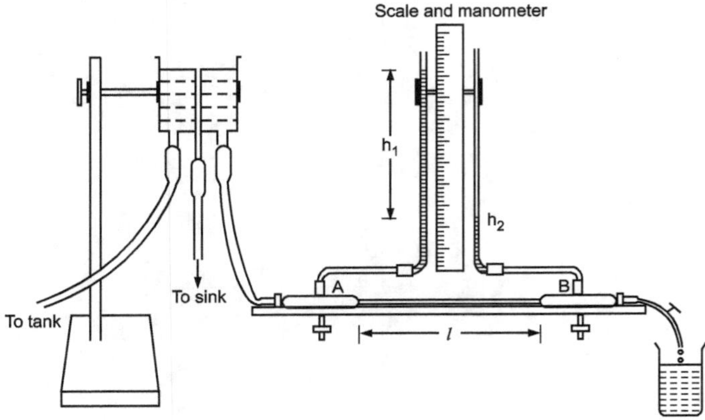

Fig. 3.1

Procedure :

1. Arrange the apparatus as shown in Fig. 3.1. Adjust the liquid flow from the tank to the constant level reservoir such that the liquid in the reservoir will always remain at a constant level.
2. Flow of water through the capillary tube is regulated by adjusting the pinch cock such that the difference of manometer levels ($h = h_1 - h_2$) remain constant. Adjust the flow such that the height difference $h = 1$ cm approximately.
3. Collect the water in a beaker for 100 sec and using a measuring cylinder, measure the volume of collected water.

4. Repeat steps (2) and (3) for height difference of 2 cm, 3 cm, 4 cm, 5 cm, and so on by adjusting the pinch cock.

5. Calculate the flow rate for various height differences, 'h'. Plot a graph of flow rate (Q) versus height difference (h) and determine its slope.

6. From the graph, determine the limiting height h_ℓ.

 **Observations :**

- Radius of capillary tube used, a = ……………..……….. cm
- Length of capillary tube, L = …………….……….. cm
- Acceleration due to gravity, g = 980 cm/sec²
- Density of water, ρ = …………………….. gm/cm³

 **Observation Table :**

Obs. No.	Manometer level (cm)		h = $(h_1 - h_2)$ (cm)	Volume 'V' of water collected in 100 seconds (cm³)				Rate of flow Q = V/100 (cm³/s)	h/Q (s/cm²)	Mean (h/Q) (s/cm²)
	h_1	h_2		(i)	(ii)	(iii)	mean			
1										
2										
3										
4										
5										
6										
7										
8										
9										
10										

 **Formulae :**

(a) Coefficient of viscosity (by calculations), $\eta = \dfrac{\pi g a^4}{8L} \times \left(\dfrac{h}{Q}\right)_{mean}$ poise

(b) Coefficient of viscosity (by graph), $\eta = \dfrac{\pi g a^4}{8L} \times \left(\dfrac{1}{slope}\right)$ poise

(c) Limiting pressure, $P_l = h_l \times \rho \times g$ dyne/cm²

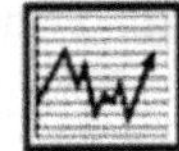

Graph :

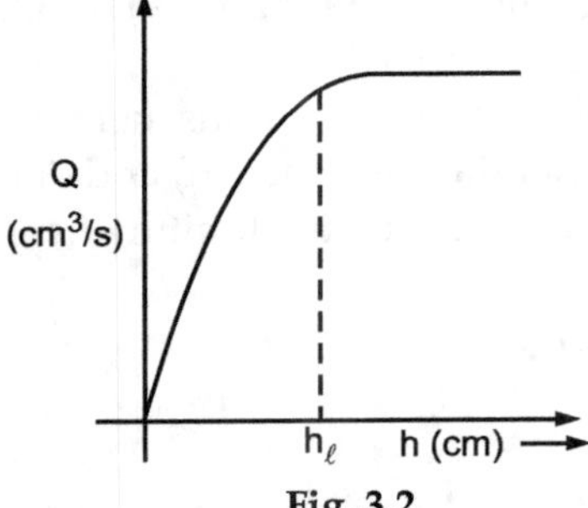

Fig. 3.2

Result :

- Coefficient of viscosity, η (by calculations) = ... poise
- Coefficient of viscosity, η (by graph) = ... poise
- Limiting pressure, P_l = ... dyne/cm²

Theory :

- Viscosity is an internal property of a fluid that offers opposition to flow.
- Fluids resist the relative motion of immersed objects through them as well as to the motion of layers with differing velocities within them.
- Formally, viscosity (represented by the symbol η "eta") is the ratio of the shearing stress (F/A) to the velocity gradient (dv_x/dz) in a fluid.
- According to Newton's formula, resulting shear of a fluid is directly proportional to the force applied and inversely proportional to its viscosity.

The Coefficient of Viscosity:

- The ratio of the shearing stress to the velocity gradient is called the coefficient of viscosity η, or $\eta = F_x/A_v$. The CGS unit for measuring the coefficient of viscosity is poise.
- The most common unit of viscosity is dyne second per square centimeter [dyne s/cm²] known as poise [P]. The SI unit of viscosity is the pascal second (Pa s).

Factors affecting viscosity:

- In general, the viscosity of a simple liquid decreases with increasing temperature (and vice versa). As temperature increases, the average speed of the molecules in a liquid increases and the amount of time they spend "in contact" with their nearest neighbours decreases. Thus, as temperature increases, the average intermolecular forces decrease.

- Viscosity is normally independent of pressure. Since liquids are normally incompressible, an increase in pressure doesn't really bring the molecules significantly close together.

Precautions :

1. Beaker for collection of water should be kept at a level lower than that of the capillary tube.
2. Ensure that height difference, h, is constant throughout the reading for a particular liquid flow rate.
3. Adjust the flow rate very carefully.
4. Do not disturb the capillary tube and the manometer throughout the experiment.

Check Your Grasp :

1. *What is meant by pressure, force and volume density?*
2. *Define: (i) Streamline flow, (ii) Turbulent flow, (iii) Critical velocity, (iv) Velocity gradient of a liquid.*
3. *State Newton's law of viscosity.*
4. *What is viscosity and coefficient of viscosity?*
5. *What are the factors affecting viscosity of a liquid?*
6. *What do you mean by a viscous medium? Give suitable examples.*
7. *Give the physical significance of limiting height and limiting pressure.*

Calculations

Y AND η BY FLAT SPIRAL SPRING

Aim : To determine Young's modulus (Y) and Modulus of rigidity (η) of material of wire by flat spiral spring.

Apparatus : Spiral spring, Vernier calliper, Screw gauge, Stop watch, Stand, Weights, etc.

Figure :

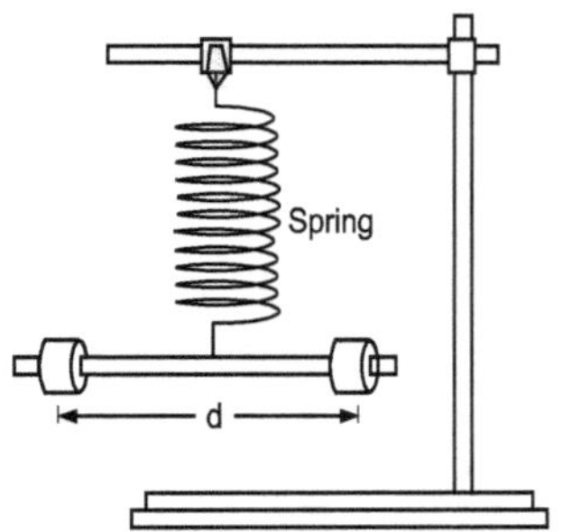

Fig. 4.1 : For measurement Young's modulus (Y)

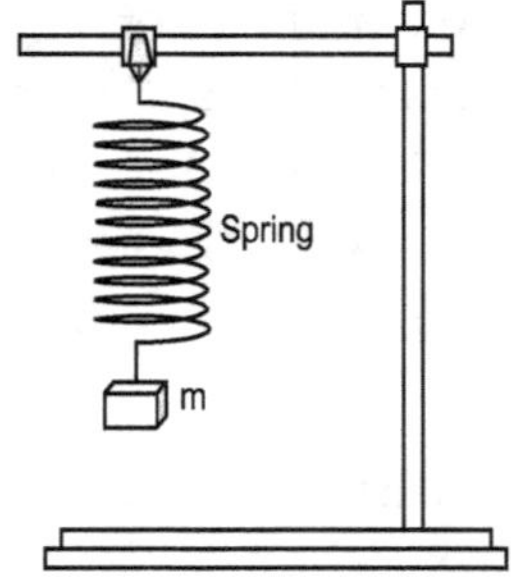

Fig. 4.2 : For measurement of Modulus of rigidity (η)

Procedure :

A. **Initial Measurements**
1. Measure the inner (D_1) and outer (D_2) diameter of the spring using vernier calliper to calculate radius of the spring (R).
2. Measure the radius of wire of the spring (r) using screw gauge.
B. **For Y**
3. Clamp the flat spiral spring firmly to a heavy retort stand. Fix the other end to a horizontal rod with two cylindrical discs placed symmetrical about centre of mass.

4. Keep a suitable distance (d) between the two discs. Give small torsional oscillations and measure the time for 20 oscillations. Determine the periodic time (T) and moment of inertia (I). Tabulate the readings.
5. Repeat step (4) for four different distances (d).

C. For η

6. Clamp the flat spiral spring firmly to a heavy retort stand and attach a suitable mass (M) to the other end.
7. Give vertical oscillations to the spring, measure time taken for 20 oscillations. Measure the periodic time (T). Tabulate the readings.
8. Repeat step (7) for three different masses (M).
9. Calculate the term $\{[M+(m/3)] /T^2\}$ in each case; and finally modulus of rigidity using its mean value.

 **Observations :**

- Least count of vernier calliper (L.C.) = ...cm
- Least count of micrometer screw gauge (L.C.) = ...cm
- Total number of turns in the spring, N = ...
- Mass of the spring, m = ... g
- Mass of the disc, m_c = ... g
- Mass of the rod, m_r = ... g
- Length of the rod, ℓ =..........cm

 **Observation Table :**

To calculate radius of spiral spring using vernier calliper

Obs. No.	Diameter	M.S.R. (a) (cm)	V.S.R. (b)	Total Reading (D) T.R. = a + (b × L.C)	Mean (cm)	Diameter of the spring	Radius of the spring
1.	inner (D_1)				$D_1 =$	D = ($D_1 + D_2$) /2	R = D/2
2.							
3.	outer (D_2)				$D_2 =$	... cm	... cm
4.							

To calculate radius of spiral spring wire using screw gauge

Obs. No.	M.S.R. (a) (cm)	V.S.R. (b)	Total reading (d') T.R. = a + (b × L.C.)	Mean d' (cm)	r = d'/2 (cm)
1					
2					

To calculate Young's modulus (Y)

Obs. No.	Distance (d) (cm)	Time for 20 oscillations (t) (s)				Period $T = \dfrac{t}{20}$ (s)	T^2 (s²)	M.I. I (gcm²)	I/T^2 (gcm²/s²)	Mean I/T^2 (gcm²/s²)
		I	II	III	Mean t					
1										
2										
3										

To calculate M.I. use $I = m_c d^2 + \dfrac{m_r l^2}{12}$

To calculate Modulus of Rigidity (η)

Obs. No.	Mass attached (M) (g)	$\left(M + \dfrac{m}{3}\right)$ (g)	Time for 20 oscillations (t) (s)				Period $T = \dfrac{t}{20}$ (s)	T^2 (s²)	$\dfrac{\left(M + \dfrac{m}{3}\right)}{T^2}$ (g/s²)	Mean $\dfrac{\left(M + \dfrac{m}{3}\right)}{T^2}$ (g/s²)
			(I)	(II)	(III)	Mean				
1										
2										
3										

$E = mc^2$ **Formulae :**

(a) Young's modulus, $Y = \dfrac{32\pi^2 RN}{r^4}\left(\dfrac{I}{T^2}\right)_{mean}$ dyne/cm²

where, R: radius of spiral spring, r: radius of the wire, N: number of turns, M: mass of each disc, $I = Md^2$: moment of inertia of the suspended bar about the axis of suspension, d: distance between the discs, T: period of **vertical** oscillations.

(b) Modulus of rigidity, $\eta = \dfrac{16\pi^2 R^3 N}{r^4}\left(\dfrac{M + \dfrac{m}{3}}{T^2}\right)_{mean}$ dyne/cm²

where, M: mass attached to the spring, R = radius of the spring + r, T : period of horizontal oscillations, m: mass of the spring.

Results :

- **Young's modulus of material of wire of the flat spiral spring,**
  Y = ... dyne/cm²
- **Modulus of rigidity of material of wire of the flat spiral spring,**
  η = ... dyne/cm²

Theory :

- Elasticity is the property of an object or material which causes it to be restored to its original shape after distortion. Elasticity is the property of solid materials to return to their original shape and size after the removal of deforming force.

- Elasticity is the amount of deformation (strain) resulting from a given stress.

- The ratio of tensile stress to tensile strain is known as Young's modulus. It can be used to predict the elongation or compression of an object.

- The SI units of Young's modulus is Pascal [Pa] or N/m².

- When the spring is suspended from the rigid support and the other end is attached to a bar with two discs as shown in Fig. 4.2 and is made to oscillate in horizontal plane, the periodic time of oscillations is given by $T = 2\pi\sqrt{\dfrac{8RNI}{r^4 Y}}$

  Therefore, the Young's modulus is $Y = \dfrac{32\pi^2 RNI}{r^4 T^2}$

  where r: radius of wire of the spring, R = radius of the spring + r, M is the applied load, N is the number of turns in the spring, I is the moment of inertia of the bar about the axis of suspension, r: radius of wire of the spring.

- In this experiment, the value of η for the material of the wire of a flat spiral spring is deduced from a knowledge of the periodic time of vertical oscillations of the spring when the dimensions of the spring and the applied load are known as they are related by the relation

$$T = 2\pi\sqrt{\dfrac{\left(M + \dfrac{m}{3}\right)2\ell R^2}{\pi r^4 \eta}}$$

Hence the modulus of rigidity is given by

$$\eta \;=\; \frac{16\pi^2 R^3 N}{r^4}\left(\frac{m + \dfrac{m}{3}}{T^2}\right)_{mean}$$

 **Precautions :**

1. Measure the radius of wire of the spring very carefully.
2. Count the oscillations from the mean position of S.H.M.
3. There should not be any kinks in the wire.
4. Amplitude of oscillations should be small.
5. Choose the spring with an appropriate value of spring constant.

 **Check Your Grasp :**

1. *Define (i) elasticity, (ii) plasticity, (iii) Young's modulus, (iv) modulus of rigidity, (v) elastic limit.*
2. *State Hooks's law in elasticity.*
3. *Explain the core concept involved in the experiment with reference to definitions of Young's modulus and modulus of rigidity.*
4. *What is the effect of temperature on Young's modulus and modulus of rigidity?*
5. *Why a flat spiral spring is used in this experiment?*
6. *Which is more elastic (a) rubber, (b) steel, (c) plastic?*

 **Calculations :**

Y BY BENDING METHOD

Aim :   To determine Young's modulus (Y) of material of a rectangular bar by bending method.

Apparatus :   Rectangular bar, Meter scale, Vernier calliper, Travelling microscope, Slotted weights, Knife edges, Hanger with pointer, Magnifying lens etc.

Figure :

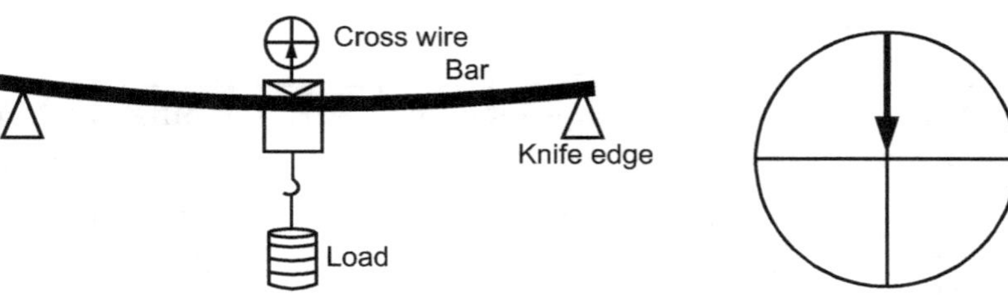

Fig. 5.1 : Bending of the bar   **Fig. 5.2 : Eye piece view**

Procedure :

1. Place the bar on two knife edges as shown in Fig. 5.1 so that the two knife edges are equidistant from C.G. (Zero position) of the bar.
2. Place the hanger and pointer system at the C.G. of the bar.
3. Level the travelling microscope and keep the telescope in horizontal plane.
4. Focus the telescope and adjust the cross wire so that tip of the pointer coincides with the horizontal line of the cross wire (Zero load reading).
5. Attach 200 g load to the hanger; adjust the vertical position of the microscope so that the tip of the pointer again coincides with the horizontal line of the cross wire.
6. Repeat the above procedure for 400 g, 600 g, 800 g, 1000 g and 1200 g (loading).

7. Decrease the load in steps of 200 g and take the corresponding readings (unloading).

8. Calculate depression of the bar (δ), plot a graph of depression versus attached load (M) and hence calculate the Young's modulus of material of the bar.

Observations :

- Least count of vernier calliper (L.C.) = ...cm
- Least count of travelling microscope (L.C.) = ...cm
- Distance between the two knife edges, ℓ =....................cm
- Width of the bar, b = ...cm.
- Breadth of the bar, d = ...cm.
- Acceleration due to gravity, g = 980 cm/s^2

Observation Table :

Obs. No.	Mass attached (M) (g)	Microscope reading (cm)			Bending $\delta = x_0 - x_n$ (cm)	M/δ (g/cm)	Mean M/δ (g/cm)
		Loading	Unloading	Mean			
1	0			$x_0 =$	---	---	
2	200			$x_1 =$	$x_0 - x_1 =$		
3	400			$x_2 =$	$x_0 - x_2 =$		
4	600			$x_3 =$	$x_0 - x_3 =$		
5	800			$x_4 =$	$x_0 - x_4 =$		
6	1000			$x_5 =$	$x_0 - x_5 =$		
7	1200			$x_6 =$	$x_0 - x_6 =$		

Formulae :

(a) Young's modulus Y (by calculations) : $Y = \dfrac{g\ell^3}{4bd^3} \times \left(\dfrac{M}{\delta}\right)_{mean}$ dyne/cm^2

(b) Young's modulus Y (by graph) : $Y = \dfrac{g\ell^3}{4bd^3} \times \left(\dfrac{1}{slope}\right)$ dyne/cm^2

where, ℓ: length, b: breadth, d: depth of the bar

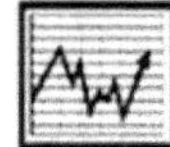

Graph :

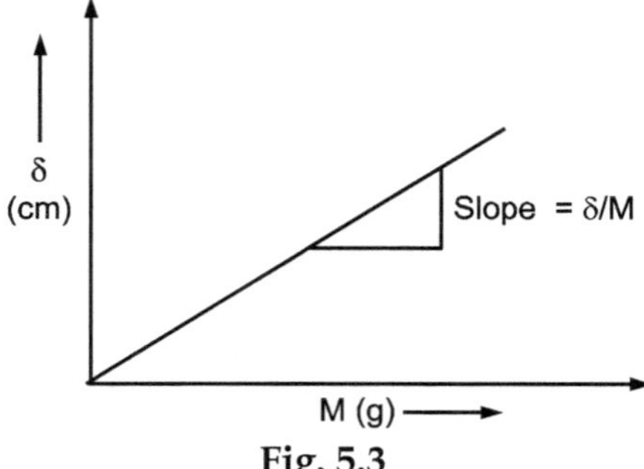

Fig. 5.3

Result :

- Young's modulus of material of the bar (by calculations),
  Y = ...dynes/cm²
- Young's modulus (by graph), Y =....................dynes/cm²

Theory :

- **Young's modulus:** It is the ratio of tensile stress to tensile strain i.e. it is the modulus of elasticity related to change in length of the object under consideration.
- There are numerous methods to determine Young's modulus. One of them is loading a rectangular bar at the centre keeping its two ends clamped. The system is also known as double cantilever. (If the bar is fixed at one end and loaded at the other, it is called a cantilever.)
- Consider a rectangular bar supported at two knife edges A and B. Let length of the bar between the two knife edges is λ in such a way that equal length of the bar project beyond the knife edges. Let the bar is loaded with a weight at the mid point, C. So, the system acts as a double cantilever.
- As the middle part of the bar is horizontal, it is equivalent to two inverted cantilevers fixed at the central point C and loaded at A and B with a load W/2 which acts vertically upwards. This produces depression, δ, at the centre; which is given by

$$\delta = \frac{(W/2)\,(\ell/2)^3}{3YI} = \frac{W\ell^3}{48YI}$$

- The moment of inertia of a rectangular bar having breadth b and thickness d is $I = \frac{bd^3}{12}$, hence the depression of the bar is given by

$$\delta = \frac{W\ell^3}{4Ybd^3} = \frac{Mg\ell^3}{4Ybd^3}$$

- Therefore, Young's modulus of material of the bar is

$$Y = \frac{Mg\ell^3}{4\delta bd^3} \quad \text{or} \quad Y = \frac{g\ell^3}{4bd^3} \times \left(\frac{M}{\delta}\right) \quad \text{dyne/cm}^2$$

 **Precautions :**

1. Knife edges should be firmly supported and be equidistant from C.G. of the bar.
2. Pointer, field of view of the telescope and eye of the observer should be in a straight line.
3. Take travelling microscope readings carefully.
4. Loading or unloading should be done gradually in equal steps.
5. Do not disturb the experimental set up throughout the experiment.

 **Check Your Grasp :**

1. *Define (i) elasticity, (ii) plasticity, (iii) modulus of elasticity, (iv) elastic limit.*
2. *What will happen if two knife edges are not equidistant from C.G. of the bar?*
3. *Define stress, strain, Y, η and K.*
4. *Explain the core concept involved in the experiment with reference to definition of Young's modulus.*
5. *If the length of the rod is made half, how will the depression change for the same applied load?*
6. *Define cantilever and bending of beam. State their applications.*

 **Calculations :**

SURFACE TENSION BY JAEGER'S METHOD

Aim :   To determine the surface tension of a given liquid by Jaeger's method.

Apparatus :   Jaeger's apparatus, Beaker, Glass tubes, Capillary, Travelling microscope, Beaker etc.

Figure :

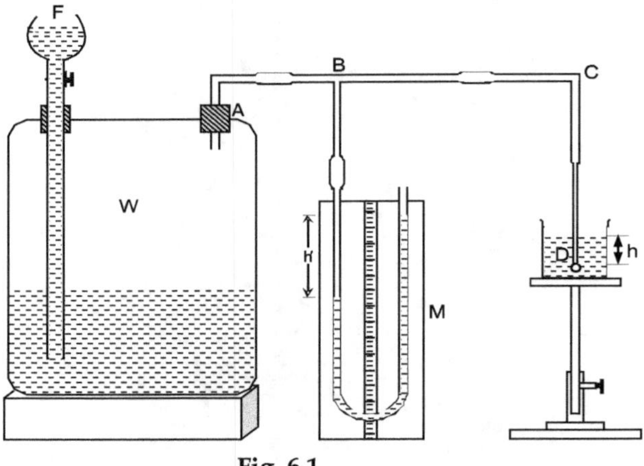

Fig. 6.1

Procedure :

1. Measure radius of the capillary tube in two mutually perpendicular positions.
2. Clamp the capillary in vertical position and place a beaker below it as shown in Fig. 6.1
3. Make a mark on the capillary tube with a permanent marker. Fill the beaker with water. Adjust the height of the beaker till the water level coincides with the mark on the capillary tube. Note the depth of the capillary dipped in the liquid (h).

4. Now, slowly open the stop cock of the stopping funnel such that water falls slowly in the bottle. This will force equal volume of air into the tube ABCD and forms bubble at the tip of the capillary, D

5. Adjust the opening of the stop cock so that a bubble is formed at D in 10 seconds. (1 bubble per 10 second)

6. When the radius of the bubble is exactly equal to that of the capillary orifice at D, pressure indicated by the manometer is maximum. Note the difference in the height (h) of liquid column in the limbs of the manometer. The bubble at this stage suddenly breaks and makes the height difference zero.

7. Repeat the experiment for thrice for the same height 'h'.

8. Repeat the complete experiment for another value of 'h'.

 **Observations :**

- Least count of travelling microscope, L.C. = ...cm
- Mean radius of the capillary orifice, r = ...cm
- Temperature of water, t = ...°C
- Density of water at t °C, ρ = ... g/cm^3.
- Density of liquid in manometer at t °C, ρ' = ... g/cm^3.

 **Observation Table :**

Table (1) : Measurement of radius of the capillary

Obs. No.	Position as seen in microscope	Microscope reading		Total reading T.R. = a + (b×L.C.)	Diameter D (cm)	Mean diameter D (cm)	Mean radius r = D/2 (cm)
		M.S.R (a) (cm)	V.S.R. (b)				
1				$R_1 =$	$D_1 = \lvert R_1 - R_2 \rvert$		
2				$R_2 =$			
3				$R_3 =$	$D_2 = \lvert R_3 - R_4 \rvert$		
4				$R_4 =$			

Table (2): Measurement of h_1

| Obs. No. | Depth of orifice (h) (cm) | Manometer levels | | | Mean (h') (cm) | Surface tension $T = \dfrac{rg\,(h'\rho' - h\rho)}{2}$ (dyne/cm) |
		Left limb reading (m) (cm)	Right limb reading (n) (cm)	Maximum difference $h' = m - n$ (cm)		
1						
2						
3						

 **Formulae :**

The surface tension of the liquid is given by,

$$T = \frac{rg\,(h'\rho' - h\rho)}{2}\ \text{dyne/cm}$$

where, r: radius of the bubble blown which is taken to be the radius of the capillary, ρ': density of the liquid present in manometer, ρ: density of the liquid in the beaker (for water, $\rho = 1$ gm/cc), h': maximum difference in the levels of the manometric liquid in the two limbs, h: depth of immersion of the capillary in the beaker, g is the acceleration due to gravity.

 **Result :**

The surface tension of the given liquid, T = ... dyne/cm

 **Theory :**

- The concept of surface tension is related to the force of cohesion between the molecules of a liquid. This force varies as the density of liquid changes. It is defined *as the force acting per unit length on the liquid on either side of the reference line taken on the liquid surface.* The direction of the force acting must be perpendicular to the selected reference line and is tangential to the surface of the liquid.
- In Jaeger's apparatus, when the capillary tube is dipped inside the liquid, the liquid level rises in it and the shape of the meniscus formed is hemispherical.

If the stop watch of the dropping funnel F is kept slightly open, the liquid (e.g. water) is allowed to fall slowly inside the bottle, W. This causes equal volume of air inside the tube ABCD; resulting in compression of the liquid (water) in the tube CD, due to which a bubble is formed at the tip of the capillary i.e. at D. The radius of the bubble formed can be varied by changing the pressure inside the tube CD. The pressure is varied till the bubble reaches to a radius 'r' which is equivalent to the radius of the capillary at 'D'.

- At this stage the manometer M shows a maximum difference in the levels of two limbs, which is indicated as h′. As the radius of bubble increases, the internal pressure gets reduced. But the pressure outside the bubble is constant which causes bubble to break.

- The pressure inside the bubble just before it breaks is given by $P + h'\rho'g$, where P: atmospheric pressure, ρ': density of liquid used in the manometer. At this instant, pressure outside the bubble will be $P + h\rho g$, where h: depth of capillary tube in the liquid, ρ : density of liquid used (e.g. water).

- Now, the excess of pressure inside the bubble is $2T/r$, we have

$$\frac{2T}{r} = (P + h'\rho'g) - (P + h\rho g)$$

$$\therefore \quad T = \frac{rg(h'\rho' - h\rho)}{2} \ \text{dyne/cm}$$

 **Precautions :**

1. Manometer should be filled with a liquid of lower density (e.g. xylol) to have a large difference between the two limbs.
2. The orifice of the capillary used should have radius of about 0.1 cm to 0.25 cm.
3. Measure the radius of the capillary tube at different positions as indicated, to eliminate possible errors.
4. Clean the glass tube before the experiment.
5. Apparatus should be air tight.
6. Keep the air space in the Woulf's bottle small so that the bubbles at the capillary tip are formed slowly and one by one. The rate of formation of the bubbles should be about one bubble per 10 seconds.

 **Check Your Grasp :**

1. *Define: (i) Adhesive force, (ii) Cohesive force, (iii) Surface tension.*
2. *What are the various methods to determine surface tension of a liquid?*
3. *What are the factors on which surface tension of a liquid depends?*
4. *Is it possible to determine surface tension of solids and gases? Justify.*
5. *What do you mean by excess of pressure?*
6. *How surface tension depends upon density of the liquid?*
7. *How shape of the meniscus of a liquid depends upon its surface tension?*

8. *State applications/ day-to-day examples of surface tension.*

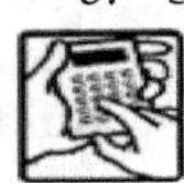 **Calculations :**

STUDY OF POISSON'S RATIO OF RUBBER USING RUBBER TUBE

Aim : To determine Poisson's ratio of rubber using rubber tube.

Apparatus : Rubber tube with metal sleeve and rubber stopper, Meter scale, Small pointer, Slotted weights of 250 gm, Hanger, Burette, Rubber stopper, etc.

Figure :

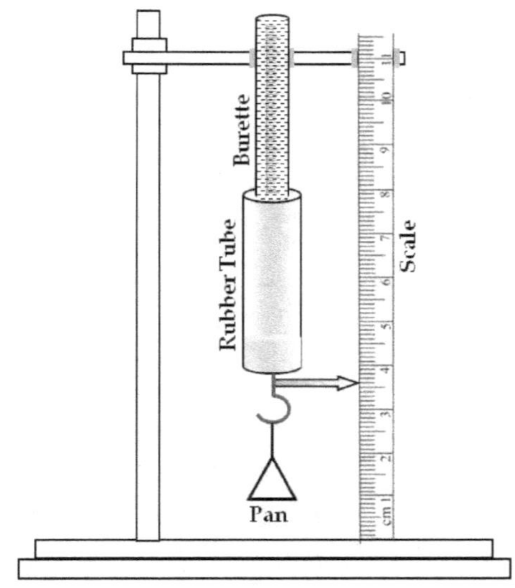

Fig. 7.1

Procedure :

1. Arrange the experimental setup as shown in the Fig. 7.1 and add pure water in the rubber tube through burette until the water meniscus appears nearly at the top of the burette.

2. Note down diameter of the rubber tube (D), position of the pointer on the scale (L) and water meniscus level in the burette (m) when no load is attached.

3. Attach a 200 gm mass at lower end of the rubber tube and note down the position on the scale (L). Also note down the water meniscus level (m) in burette for this mass.

4. Add mass in steps of 200 gm (loading) and note down the corresponding pointer position and burette water meniscus reading.

5. Repeat the above procedure for unloading; that is, by decreasing the weights in steps of 200 gm and note down the corresponding values of L and m.

6. Complete the observation table and accordingly calculate dV/dL.

Observations :

- Least count of vernier calliper , L.C. = ...cm
- Diameter of rubber tube without load, (D)

  D_1 = ...cm D_2 = ...cm D_3 = ...cm.

  Mean diameter $D = \dfrac{D_1 + D_2 + D_3}{3}$ = ...cm.

- Radius of the rubber tube, $r = \dfrac{D}{2}$ = ...cm.

- Cross sectional area of the rubber tube, $A = \pi r^2$ = ...cm^2.

- Inner diameter of the burette, (d)

  d_1 = ...cm d_2 = ...cm d_3 = ...cm.

  Mean diameter $d = \dfrac{d_1 + d_2 + d_3}{3}$ = ...cm.

- Inner radius of the burette, $r' = \dfrac{d}{2}$ = ...cm.

- Cross sectional area of the burette, $a = \pi r'^2$ = ...cm^2.'

 **Observation Table :**

Obs. No.	M (gm)	Reading of pointer on scale L (cm)			dL (cm)	Water meniscus reading on burette 'm'			Volume V = a × m (cm³)	dV (cm³)	dV/dL
		Load X	Unload Y	Mean $L = \dfrac{X + Y}{2}$		Load X′	Unload Y′	Mean $m = \dfrac{X' + Y'}{2}$			
1	0										
2	200										
3	400										
4	600										
5	800										
6	1000										
7	1200										
										Mean	

M : Mass attached, dL: Change in length and and dV: Change in Volume

 **Formulae :**

The Poisson's ratio of rubber tube is given by,

$$\sigma = \frac{1}{2}\left[1 + \frac{1}{A}\left(\frac{dV}{dL}\right)_{mean}\right]$$

$$\sigma = \frac{1}{2}\left(1 + \frac{1}{A}\,\text{Slope}\right)$$

where, A: cross sectional area of rubber tube.

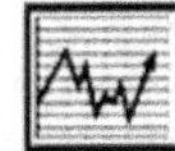

Graph :

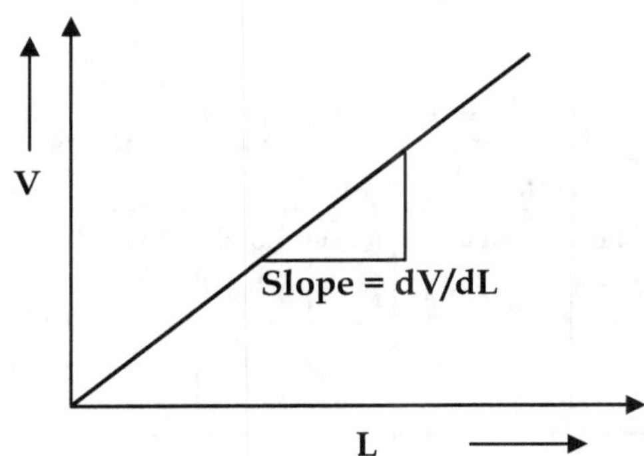

Result :

The Poisson's ratio of the given rubber tube, σ = ...

Theory :

- **Elasticity** is the property of materials due to which materials can regain their original shape and size after the removal of deforming forces.
- **Strain** is defined as the change in length divided by the original length.
- **Stress** is defined as the average restoring force per unit area.
- **Elastic limit** is the maximum stress or force per unit area within a solid material that can arise before the permanent deformation.
- **Hooke's law** states that, within the elastic limit, stress is directly proportional to strain.
- **Young's modulus** is the ratio of tensile or compressive stress to tensile strain.
- **Modulus of rigidity** is the ratio of shearing stress to shearing strain.
- **Bulk Modulus** is the ratio of volumetric stress to the volumetric strain.
- Extension and contraction are opposite types of linear strain. Whenever a material is extended or contracted by a linear stress in one direction, the reverse strain usually takes place in the perpendicular directions. Direction of linear stress is called the axial direction. All the directions that are perpendicular to this are called the transverse directions.
- Ratio of lateral strain to longitudinal strain is called as Poisson's ratio.

Precautions :

1. While filling up the rubber tube and burette, care should be taken that there should be no air bubble inside it.
2. Mass should be attached or removed in equal steps.
3. Reading should be taken when water meniscus level and pointer position are stationary.
4. Mass suspended at the lower end of the rubber tube should not exceed the maximum mass permissible within the elastic limit.

Check Your Grasp :

1. *Define: (i) Young's modulus, (ii) Modulus of rigidity, (iii) Elastic limit.*
2. *Define the term: Poisson's ratio.*
3. *State Hooke's law in elasticity.*
4. *What do you mean by elasticity?*
5. *What is the significance of Poisson's ratio?*
6. *What do you mean by lateral strain and longitudinal strain?*

Calculations :

SURFACE TENSION OF LIQUID USING FERGUSSON METHOD

Aim :   To determine the surface tension of a given liquid (water) by Fergusson method

Apparatus :   Manometer, Travelling Microscope, Capillary tube, etc.

Figure :

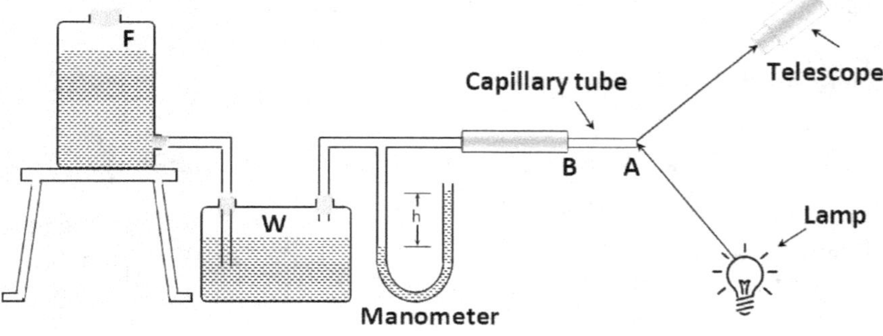

Fig. : 8.1

Procedure :

1. Determine radius of the capillary tube using a travelling microscope.
2. Introduce a small quantity of water in capillary tube using a rubber dropper as shown in Fig. 8.1.
3. Focus light on the capillary tube (end A) and focus the telescope at point A.
4. Water reservoir (F) is adjusted at such a level that the water thread in the capillary moves towards open end, A.

5. Adjust the level of 'F' till the liquid at 'A' forms a flat surface at the end of the tube. (Initially, when the liquid forms a plane surface at 'A', it appears to be uniformly illuminated through the telescope; else before adjusting the level of 'F', image of the lamp filament is seen through the telescope. However, when level of 'F' is adjusted slowly, the filament image (in the telescope) gradually becomes broader and the moment liquid in the capillary forms a flat surface at 'A', the filament image disappears and telescope shows a bright, well illuminated spot.

6. Note down the heights h_1 and h_2 of the manometer when the filament image disappears (or a bright spot is observed through the telescope).

7. Adjust the position of water reservoir (F) few more times so as to get the critical position where the liquid surface just attains flatness. This gives accuracy in surface tension to be calculated.

8. Calculate the surface tension of the liquid using the given formula.

Observations :

- Least count of travelling microscope, L.C. = ...cm
- Radius of capillary tube, a = ...cm.
- Density of liquid in manometer at t °C, ρ = ... g/cm³.

Observation Table :

Table (1) : Measurement of radius of the capillary

Obs. No.	Position as seen in microscope	Microscope reading		Total reading T.R. = a + (b×L.C.)	Diameter D (cm)	Mean diameter D (cm)	Mean radius r = D/2 (cm)
		M.S.R (a) (cm)	V.S.R. (b)				
1				$R_1 =$	$D_1 = |R_1 - R_2|$		
2				$R_2 =$			
3				$R_3 =$	$D_2 = |R_3 - R_4|$		
4				$R_4 =$			

Table (2): Measurement of h_1 and h_2

Obs. No.	Manometer levels			Mean (h) (cm)	Surface tension $T = \dfrac{h\rho gr}{2}$ (dyne/cm)
	h_1 (cm)	h_2 (cm)	Difference $h = h_1 - h_2$ (cm)		
1					
2					
3					

Formula :

The surface tension of the liquid is given by,

$$T = \frac{h\rho gr}{2}$$

where, ρ is density of the liquid and g is the acceleration due to gravity.

Result :

Surface tension of the given liquid, T = ... dyne/cm

Theory :

- Concept of surface tension is related to the force of cohesion between the molecules of a liquid. This force varies as the density of liquid changes. It is defined *as the force acting per unit length on the liquid on either side of the reference line taken on the liquid surface.* Direction of the force must be perpendicular to the selected reference line and is tangential to the surface of the liquid.
- **Surface tension:** The force acting perpendicular to unit length of an imaginary line drawn in the plane of the liquid surface is called surface tension of that liquid.
- Molecules located on the surface of a liquid possess potential energy and as the potential energy of molecule tends to have a minimum value, the liquid tries to have the least possible number of molecule on its surface, by reducing its surface area; causing surface tension.

- Surface tension is caused by the inward attraction on the molecules on the surface. The attraction produces curvature in free liquid surfaces and causes a pressure difference to exit at the curved boundary.
- In C.G.S system, surface tension is expressed in dynes/cm.
- Surface tension depends on factors as (i) nature of the liquid (ii) nature of surfaces in contact (iii) temperature of liquid.
- Temperature of the liquid significantly affects the surface tension. At the higher temperature, surface tension decreases and vanishes at the critical temperature.
- Molecules in the liquid state experience strong intermolecular attractive forces. Force of attraction between molecules of the same substance is called cohesive force and the force of attraction between molecules of different substances is called an adhesive force. The strong cohesive forces constitute surface tension of liquid. The strong adhesive force causes capillary action.
- **Angle of contact:** When a liquid is in contact with a solid, angle between the surface of the solid and the tangent drawn to the free surface of liquid; at the point of contact, measured from inside the liquid is called the angle of contact.

 **Precautions :**

1. Measure radius of the capillary tube at different positions using travelling microscope as indicated, to eliminate possible errors.
2. Dip the capillary tube in the water carefully and check concave shape of water in the tube.
3. Clean the capillary tube before the experiment.
4. Apparatus should be air tight.
5. Focus the light on the capillary tube carefully

 **Check Your Grasp :**

1. *Define: (i) Adhesive force, (ii) Cohesive force,(iii) Surface tension.*
2. *What are the various methods to determine surface tension of a liquid?*
3. *What are the factors on which surface tension of a liquid depends?*
4. *Is it possible to determine surface tension of solids and gases? Justify.*
5. *What do you mean by excess of pressure?*
6. *How surface tension depends upon density of the liquid?*
7. *How shape of the meniscus of a liquid depends upon its surface tension?*
8. *State applications/ day-to-day examples of surface tension.*

 **Calculations :**

❖ ❖ ❖

II – 1

/ /

Experiment performed on

SPECTROMETER & DETERMINATION OF ANGLE OF PRISM

Aim : To study the optical instrument - spectrometer and to determine the angle of a given prism.

Apparatus : Spectrometer, Mercury lamp, Prism, Reading lamp, Spirit level, Magnifying lens etc.

Figure :

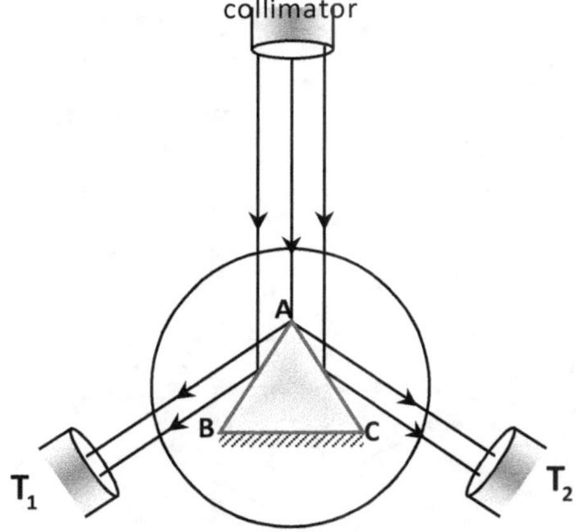

Fig. 1.1

Procedure :

1.   Level the spectrometer base, prism table, telescope and collimator.
2.   Move the telescope to make collimator and telescope coaxial. Observe an image of the slit through the telescope. Adjust the slit width to make the slit narrow.
3.   Adjust the spectrometer for parallel rays of light using Schuster's method.

4. Keep the prism at the centre of prism table in such a way that the rough surface of the prism is perpendicular to the collimator axis as shown in Fig. 1.1.
5. Move the telescope towards left (position T_1) till the image of the slit (reflected from face AB) is seen. Note the total reading from window 1 and window 2.
6. Move the telescope towards right (position T_2) till the image of the slit (reflected from face AC) is seen. Note the total reading from window 1 and window 2.
7. Tabulate the readings in the table and calculate the angle of prism, A.

 **Observations :**

- Smallest division on main scale = ……………… degree.
- Total number of divisions on vernier scale =………………

$$\text{Least count of spectrometer, LC} = \frac{\text{Value of smallest division on main scale}}{\text{Total number of divisions on vernier scale}}$$

$$= \ \text{…………… min.}$$

 **Observation Table :**

Telescope on L.H.S. (Position T_1)		Telescope on R.H.S. (Position T_2)		$2A = $ $\lvert \theta_1 - \theta_2 \rvert$ degree (x)	$2A' = $ $\lvert \theta_1' - \theta_2' \rvert$ degree (y)	Angle of prism (A) degree (x + y)/4
Window 1 θ_1 degree	Window 2 θ_1' degree	Window 1 θ_2 degree	Window 2 θ_2' degree			

 **Result :**

Angle of the given prism, A = ……………………. degree

 **Theory :**

Spectrum:
- It is obtained on dispersion of light through a prism as refractive index, μ, of a prism is different for different colours i.e. for different wavelengths.

- There are three types of spectra: (a) Continuous spectra, (b) Band spectra and (c) Line spectra. To obtain a particular type of spectrum, certain conditions are to be satisfied.

Spectrometer :

- The spectrometer is an instrument for analyzing the spectra of radiations. Various forms of the spectrometer are used for different parts of the electromagnetic spectrum and for different purposes.
- A spectrometer has three main parts viz. a collimator, telescope and prism table.
- A spectrometer is to be levelled using a spirit level, initially so as to have the spectrometer and its prism table exactly horizontal. Normally collimator and telescope are levelled by manufacturer for parallel axes.
- Before using the spectrometer, it should be adjusted by Schuster's method for the best adjustment of the collimator and telescope.
- Schuster's method is the method of adjusting the collimator and telescope for parallel ray of light. With these adjustments, the rays entering and leaving the prism are parallel.

About the experiment:

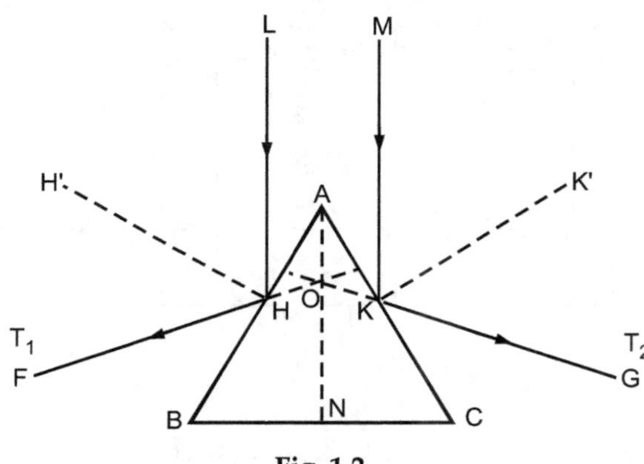

Fig. 1.2

In Fig. 1.2, MK is the incident ray at point K.

∴ $\angle MKK' = \angle K'KG$ (angle of incidence = angle of reflection)

∴ $\angle MKA = \angle GKC$ (opposite angles)   ... (1)

Also, $\angle MKA = \angle KAO$ (alternate angles) and $\angle GKC = \angle AKO$ (opposite angles)

∴ $\angle MKA = \angle GKC = \angle KAO = \angle AKO = \dfrac{A}{2}$ ... (2)

Now, $\angle KON$ is an exterior angle of ΔKAO

$$\therefore \quad \angle KAO + \angle AKO = \angle KON \Rightarrow \angle KON = \frac{A}{2} + \frac{A}{2} = A \qquad \ldots (3)$$

Similarly, it can be shown that $\angle HON = A$

$$\frac{1}{2}(\angle KON + \angle HON) = \angle A$$

Hence, the angle between the direct position and position T_1 of the telescope is A. Similarly angle between direct position and position T_2 of the telescope is also A.

While performing the experiment, angle between the positions T_1 and T_2 is measured as 2A from which angle of prism A is calculated.

Precautions :

1. Level the spectrometer accurately using spirit level bottle before starting the experiment.
2. Slit should be as narrow as possible.
3. Adjust the eye piece of the telescope so as to observe the vertical cross wire to be exactly parallel to the image of the slit.
4. Prism should be placed at the centre of the prism table.

Check Your Grasp :

1. *What type of source of light you are using? Why?*
2. *What are the types of spectra? What are the conditions to obtain them?*
3. *What is a spectrometer?*
4. *Explain use of each part of a spectrometer.*
5. *Why is it necessary to adjust the spectrometer for parallel rays of light?*
6. *What are different types of lenses?*
7. *What types of lenses are used in a spectrometer?*
8. *What is reflection and refraction?*
9. *What do you mean by a prism? Explain the terms related to a prism.*
10. *What are the types of a prism?*
11. *Why a spectrum is not observed in this experiment?*

 **Calculations :**

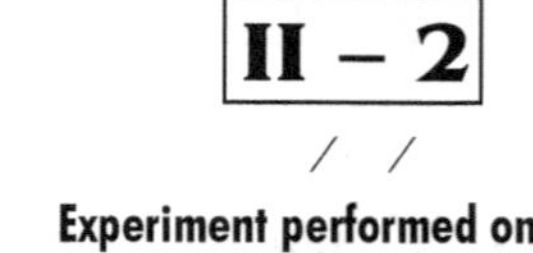

CALIBRATION OF A SPECTROMETER

Aim : To calibrate the spectrometer for parallel ray of light. Also, determine refractive index of different colors and find out wavelength of unknown colour of light.

Apparatus : Spectrometer, Mercury lamp, Prism, Reading lamp, Spirit level, Magnifying lens, etc

Figure :

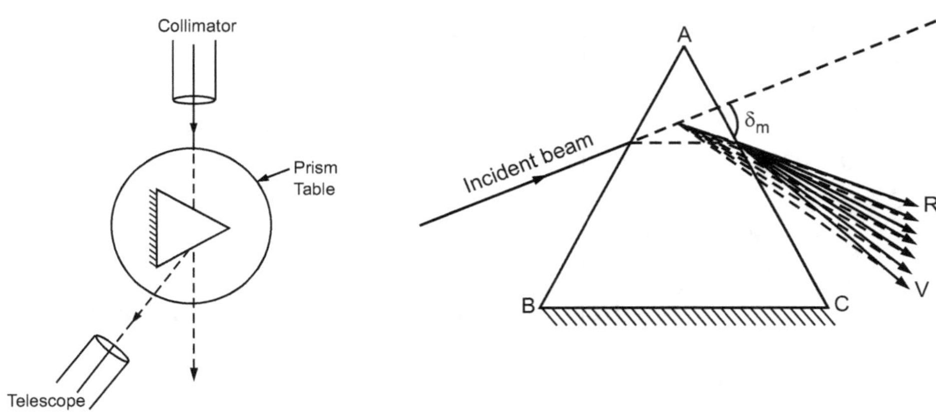

Fig. 2.1   Fig. 2.2

Procedure :

1. Level the spectrometer base, prism table, telescope and collimator.
2. Make collimator and telescope coaxial. Observe image of the slit through the telescope. If needed, adjust the slit width to make the slit narrow.
3. Keep the prism on the prism table such that one of the refracting surfaces faces the collimator as shown in the figure. A spectrum will be seen through the other surface of the prism.
4. Move the telescope, focus it on the spectrum.
5. Adjust the spectrometer for parallel rays of light using Schuster's method.

6. Adjust the spectrum for minimum deviation position by rotating the prism table. Coincide the vertical cross wire of the telescope with violet colored spectral line. Note down the reading of the spectrometer from any one window.
7. Repeat the process for other distinct colors of the spectrum.
8. Remove the prism and focus the telescope on the direct slit image.
9. Note down the 'direct reading' from the same window of the spectrometer.
10. Determine the values of refractive index for different colors.
11. Plot the calibration curves of wavelength (λ) versus refractive index (μ) and also between ($1/\lambda$) versus mean (δ) as shown.
12. Determine the wavelength of known colored spectral line using the calibration curve.

 **Observations :**

- Angle of prism, $A = 60°$
- Smallest division on main scale = ... min
- Total number of divisions on vernier scale = ...
- L.C. of spectrometer = $\dfrac{\text{Value of smallest division on main scale}}{\text{Total number of divisions on vernier scale}}$ = ...min
- Direct reading of the spectrometer, V_A = ..., V_B = ...

 **Observation Table :**

| Colour | Wave-length (λ) Å | Spectrometer reading | | Angle of minimum deviation | | | $1/\lambda^2$ | μ |
		V'_A (deg)	V'_B (deg)	$\delta_1 = V_A - V'_A$ (deg)	$\delta_2 = V_B - V'_B$ (deg)	Mean (δ_m) (deg)		
Violet								
Blue								
Green								
Yellow-I								
Yellow-II								
Unknown								

 **Formulae :**

Refractive index of the prism for each colour using the formula,

$$\mu = \frac{\sin\left(\dfrac{A + \delta_m}{2}\right)}{\sin\left(\dfrac{A}{2}\right)}$$

where, δ_m is the angle of mean deviation for the particular colour.

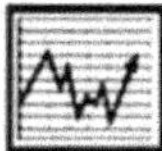 **Graphs :**

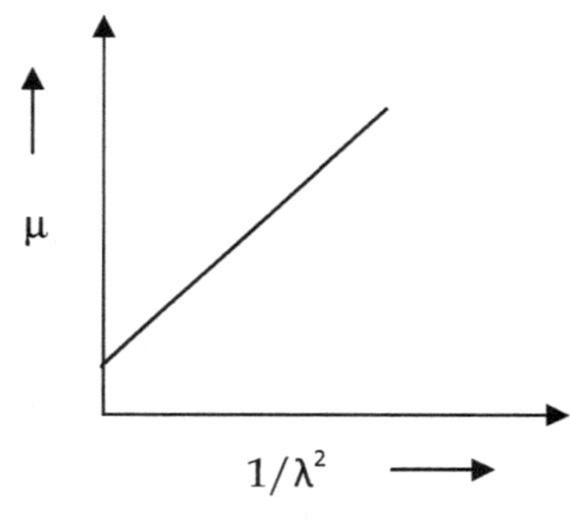

Fig. 2.3

 **Results :**

- The calibration curves for the given prism is plotted.
- Refractive indices of different colours were determined and are tabulated in the observation table.
- The observed wavelength of unknown colored spectral line using calibration curve is…………. which corresponds to …………..colored spectral line.

 **Theory :**

- When a ray of light is incident on a prism, it gets refracted through it and hence gets deviated from its original path.
- The angle between the direction of the incident ray and the emergent ray is called angle of deviation.
- Such deviation depends upon angle of incidence and the wavelength of the light used. For a certain angle of incidence, angle of deviation is minimum and is denoted by δ_m. The relation between angle of incidence; i, angle of minimum

deviation; δ_m, angle of prism; A and refractive index of the prism (for the incident wavelength λ) is given by the prism formula as:

$$\mu = \frac{\sin\left(\frac{A + \delta_m}{2}\right)}{\sin\left(\frac{A}{2}\right)}$$

- As angle of deviation, δ, is inversely proportional to the square of the wavelength of the light, the graph between $1/\lambda$ and δ_m is a straight line. From this plot, unknown wavelength can be calculated. This graph is called calibration curve.

Precautions :

1. Slit should be as narrow as possible.
2. Adjust the eye piece of the telescope so as to observe the cross wire without any strain on eyes.
3. Schuster's method should be carried out with at most care and accuracy.
4. Distance of source from the slit should remain constant throughout the experiment.
5. Spectrometer scale should be viewed normally.

Check Your Grasp :

1. *What type of source of light you are using? Why?*
2. *What are the types of spectra? What are the conditions to obtain them?*
3. *What is a spectrometer?*
4. *Explain use of each part of a spectrometer.*
5. *Why is it necessary to adjust the spectrometer for parallel rays of light?*
6. *What are different types of lenses?*
7. *What do you mean by a prism? Explain the terms related with a prism.*
8. *What are the types of a prism?*
9. *What is meant by minimum deviation?*
10. *What is the practical importance of the minimum deviation position of the prism?*
11. *Why refractive index material of prism varies with colour of incident light?*
12. *What is the necessity of calibrating the given prism?*
13. *How unknown wavelength is calculated using the calibration curve?*
14. *What are the limitations on use of calibration curve?*
15. *How the wavelength and angle of minimum deviation are related?*

Calculations :

DIVERGENCE OF LASER BEAM

 **Aim :**   To study the divergence of laser beam.

 **Apparatus :** LASER source, two screens, meter scale, travelling microscope, etc

 **Figure :**

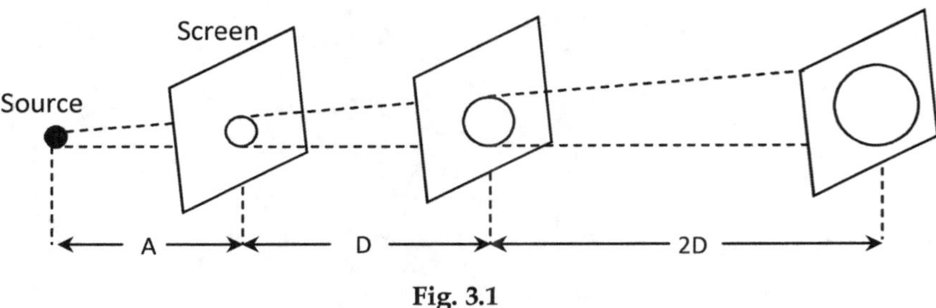

Fig. 3.1

 **Procedure :**

1. Adjust the LASER such that its beam is parallel to the earth's surface.
2. Keep three different positions of the screen at which the beam spot is measured i.e. A, A+D and A+2D respectively.
3. The distance between LASER source and the third position of the screen must be greater than at least six meters.
4. Mark the spot at these three positions.
5. Measure the diameter of each spot using travelling microscope.
6. Calculate the divergence of the beam by using the formula given as above

 **Observations :**

- Least count of travelling microscope = ……..cm
- Distance D between Ist and IInd screen is = ……..cm

 **Observation Table :**

Sr. No.	Position of screen from LASER gun	Position of T.M	M.S.R a (cm)	V.S.R b	V.S.R x LC c (cm)	a+c (cm)	Diameter (cm)
1	A = ….					x_1	$W_1 = \mid x_1 - y_1 \mid$ = ……
						y_1	
2	A+D = ….					x_2	$W_2 = \mid x_2 - y_2 \mid$ = ……
						y_2	
3	A+2D = ….					x_3	$W_3 = \mid x_3 - y_3 \mid$ = ……
						y_3	

 **Formulae :**

The divergence of laser beam is given by,

$$\theta = \frac{1}{D} \sqrt{\frac{W_3^2 - 2W_2^2 + W_1^2}{2}}$$

where, D is the distance between 1st position of the screen and 2nd position of the screen, W_1 is the diameter of spot at distance A, W_2 is the diameter of the spot at distance A+D and W_3 is the diameter of the spot at a distance A+2D.

Result :

Divergence of the laser beam, θ_0 = ...rad.

Theory :

- **Laser**: It is the acronym for Light Amplification by Stimulated Emission of Radiation.

- **Induced absorption**: When an incident photon is absorbed by an atom in ground state (E_1), it goes into excited state (E_2). This transaction is known as induced absorption.

- **Spontaneous emission**: Electron from the excited state decays into lower energy state with the emission of photon. This transaction is known as spontaneous emission.

- **Stimulated emission**: The process of emission of photon by excited atom through a forced transition occurring under the influence of external agent is known as stimulated emission.

- **Metastable state**: The long lived, intermediate energy state is called as metastable state.

- **Characteristics of Laser**

    1. **Coherence**: In laser light, a large number of identical photons are emitted through stimulated emission; therefore they are in the same phase with each other and the emitted laser light is highly coherent.

    2. **Directionality**: In a laser beam, all the photons travel in the same direction. Therefore, light emitted by a laser is unidirectional.

3. **Monochromatic**: The photons are emitted via stimulated emission; hence they possess the same energy and thus have a single wavelength and are monochromatic in nature.

4. **High Intensity**: Intensity of laser light is high and it does not vary with distance.

5. **Divergence**: Divergence of the laser beam is defined as magnitude of spread of the beam with distance. It is expressed in terms of the angle (in radian) subtended by the laser spot to the center of laser source.

6. From experimental setup discussed in earlier sections, the spot diameters are

$$W_1 = x \times \theta,$$
$$W_2 = (x + D) \times \theta$$

and
$$W_3 = (x + 2d) \times \theta$$

Thus,
$$W_1{}^2 + W_3{}^2 - 2W_2{}^2 = (x \times \theta)^2 + ((x + 2D) \times \theta)^2 - 2\,[(x + D) \times \theta]^2.$$

On simplification we get, $W_1^2 + W_3^2 - 2W_2^2 = 2D^2\theta^2$

Therefore, the divergence of laser beam is given by, $\theta = \dfrac{1}{D}\sqrt{\dfrac{W_3^2 - 2W_2^2 + W_1^2}{2}}$

Precautions :

1. Screen should be exactly perpendicular to the horizontal laser beam.

2. Laser source must be horizontal throughout the experiment.

3. Initial spot size (W_1) should be large.

Check Your Grasp :

1. *What do you mean by divergence of laser?*

2. *State the characteristics of laser.*

3. *What is the significance of low and high divergent light source?*

4. *What are the practical uses of a laser?*

5. *Define the terms: Stimulated emission and Spontaneous emission?*

6. *What is a metastable state? What is its significance in laser light?*

 **Calculations :**

❖ ❖ ❖

STUDY OF TOTAL INTERNAL REFLECTION USING LASER

Aim : To determine refractive index of water by means of total internal reflection using LASER light.

Apparatus : LASER source, Spectrometer, Screen with a stand, Rectangular trough system, Spirit level, Magnifying lens, etc.

Figure :

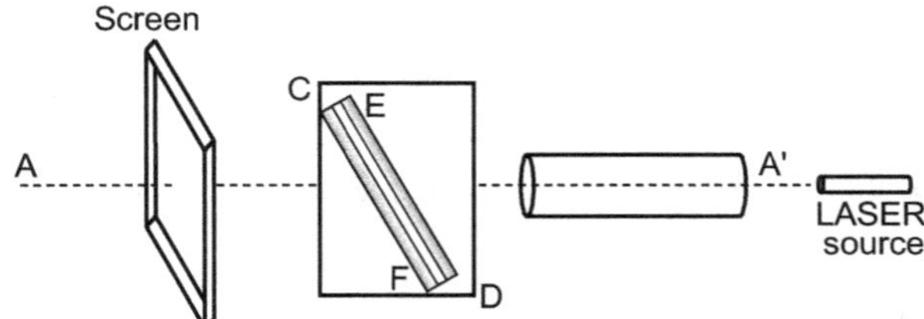

Fig. 4.1

Procedure :

1. Arrange the LASER source, spectrometer, rectangular trough system CD and the collimator such that the LASER beam is parallel to the collimator axis. Fill the container with the liquid (e.g. water) whose refractive index is to be determined.

2. Initially, no light will be produced on the screen.

3. Now, rotate the trough till a spot just appears on the screen. Note down the position of the trough on the spectrometer scale.

4. Turn the trough from this position till the image again disappears. Note down this position of the trough.

5. Calculate the total angle of rotation. Angle of incidence is half the angle of rotation. Using the formula; calculate the refractive index of the liquid.

Observation Table :

Obs. No.	Initial position			Final position			Total Angle $2\theta = |\theta_2 - \theta_1|$ (deg.)	Angle of Incidence (θ) (deg.)
	C.S.R. (deg.)	V.S.R.	Total (θ_1) (deg.)	C.S.R. (deg.)	V.S.R.	Total (θ_2) (deg.)		

Formulae :

$$_a\mu_w = \frac{1}{\sin\theta}$$

$_a\mu_w$ is the refractive index of water w.r.t air, θ is the angle of incidence.

Result :

Refractive index of water w.r.t air, $_a\mu_w$ =.........................

Theory :

About the apparatus:

The experiment requires a trough system as shown in Fig. 4.2. It contains a small rectangular air cell EF which is made up of two parallel glass plates cemented together with a thin air space between them.

Fig. 4.2

Total Internal Reflection:

- When a ray of light travels from a denser medium to rarer medium and is incident on the interface at an angle greater than a particular angle (critical angle), the ray gets totally reflected back. This phenomenon is called total internal reflection.

- **Conditions for TIR:** (i) Beam of light should be incident on a rarer medium from denser medium. (ii) Angle of incidence should be equal to or greater than the critical angle for the pair of the media.
- When the angle of incidence at the glass surface is equal to the critical angle, the incident light is totally reflected so that no refraction takes place and the spot received on the screen disappears.
- When a ray of light passes from a medium of refractive index μ_1 to a medium with refractive index μ_2 at an angle of incidence I and gets refracted at an angle of refraction, r, we have:

$$_1\mu_2 = \frac{\mu_2}{\mu_1} = \frac{\sin i}{\sin r} \qquad \dots (1)$$

About the experiment:

- Consider a beam of light incident from water (through the container CD filled with water) on the glass plate and gets totally reflected at the glass air interface at the critical angle as shown in Fig. 4.3.
- Let $_a\mu_g$ is the refractive index of glass w.r.t. air, FBG, CH and F′DG′ represent the normals to the glass surface at B, C and D respectively, then we have,

$$\frac{1}{_a\mu_g} = \sin \angle BCA \qquad \dots (2)$$

Also,

$$_w\mu_g = \frac{\sin ABF}{\sin CBG} = \frac{\sin ABF}{\sin BCH'} \quad \dots (3)$$

If the incident light gets totally reflected, at point C, $\angle BCH = 90°$, hence $\sin BCH = 1$

$$_w\mu_g = \frac{1}{\sin ABF} \qquad \dots (4)$$

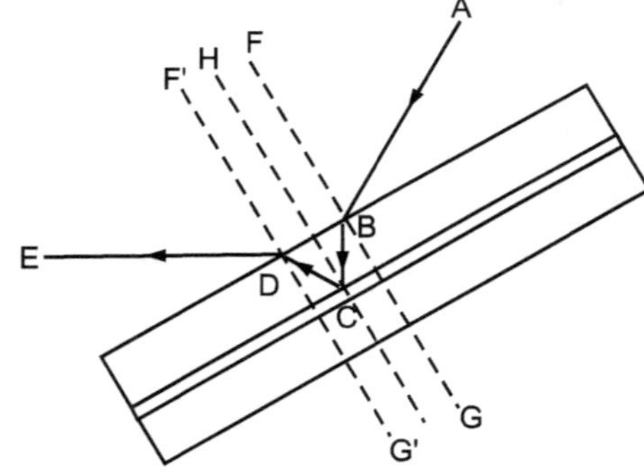

Fig. 4.3

- Thus, if angle ABF is measured experimentally, refractive index of water (the liquid under consideration) can be obtained.
- Consider Fig. 4.4. Let AA′ denote the line along which the LASER beam emerges out from the trough and EF – the first position when light just appears. The second position, when the beam disappears will be E′F′ if $\angle AOE = \angle AOE'$. This second condition can be obtained by turning the trough through either $\angle EOE$ or $\angle EOF'$. From the geometry, $\angle EOF' = \pi - \angle EOE'$

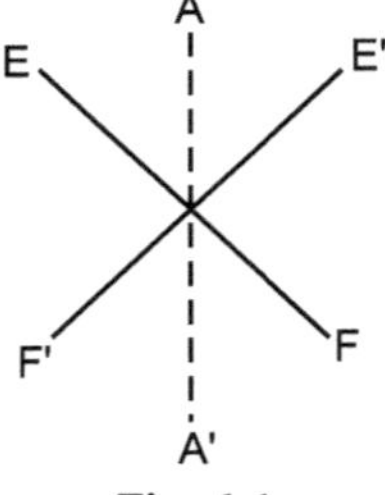

Fig. 4.4

- To determine the refractive index of water here, angle between the ray and the normal to EF or E′F′ i.e. the angle between OA and the normal to EF or E′F′ is required. From the geometry, it can be seen that this required angle is half the angle between the two normals.
- Hence, if the angle through which the air cell EF rotates is measured, the refractive index of the liquid under consideration can be obtained.

Precautions :

1. Do NOT look into the LASER source; it can be hazards to eyes.
2. Measure the least count of the spectrometer carefully.
3. Rotate the trough system precisely.
4. Ensure proper alignment of all the components before the experiment starts.

Check Your Grasp :

1. *What does LASER stands for?*
2. *What do you meant by total internal reflection?*
3. *State the laws of reflection and refraction.*
4. *How total internal reflection differs from regular reflection?*
5. *What is the basic cause of TIR?*
6. *What are the factors that influence TIR?*
7. *Define: (i) Absolute refractive index, (ii) Relative refractive index, (iii) Critical angle.*
8. *Why a LASER beam is used in this experiment instead of other monochromatic source of light?*
9. *How the result will affect if the LASER source is replaced by sun light?*
10. *State three applications of total internal reflection.*

Calculations :

DETERMINATION OF PLANCK'S CONSTANT

 **Aim :** To determine Planck's constant using a photocell.

 **Apparatus :** Photocell enclosed in a blackened wooden box, Set of three filters (Red, Green, Blue), Incandescent bulb, DC voltmeter, DC micrometer, connecting wires etc.

 **Figure :**

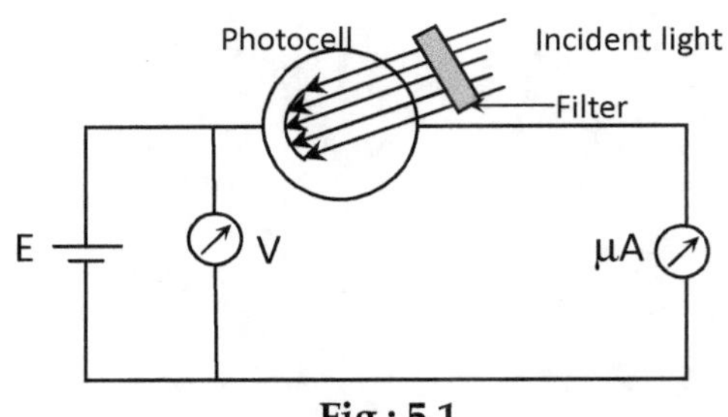

Fig : 5.1

 **Procedure :**

1. Make the connections as shown in Fig.5.1.
2. Set micro-ammeter reading to zero, when no light falls on the photocell. For this the window of photocell should be closed by a cap so that even no stray light falls on the photocell. 'Zero current setting' knob is also provided on certain experimental set-ups.
3. Now, the open the photocell window and turn ON the light source. This causes some definite current to flow through the circuit.

64

4. Place a given filter (Red) of known wavelength ($\lambda_{1)}$ in the path of the light by putting it at the window of photocell. Some current will be observed in the ammeter. This current corresponds to zero anode potential.

5. Increase the negative anode potential gradually in steps and note down the ammeter reading in each case till the current reduces to zero (stopping potential for λ_1).

6. Note down the voltage (V) corresponding to the zero current. This is the stopping voltage (V_R) for a red filter of wavelength λ_1.

7. Replace the red filter by green and then by blue. Measure the stopping voltage V_G, V_B by same procedure.

Observation Table :

Obs No	Stopping potential for (in volt):		
	Red filter (λ_R = 6560 Å)	Green filter (λ_R = 5460 Å)	Blue filter (λ_R = 4360 Å)
1	$V_R=$	$V_G=$	$V_B=$
2	$V_R=$	$V_G=$	$V_B=$
3	$V_R=$	$V_G=$	$V_B=$
Mean	$V_{R(mean)} =$	$V_{G(mean)} =$	$V_{B(mean)} =$

Formulae :

$$h_1 = \frac{e}{c} \times \left(\begin{array}{c} \text{slope of the plot} \\ \text{between Green \& Red} \end{array} \right) \times \lambda_R \lambda_G \qquad h_2 = \frac{e}{c} \times \left(\begin{array}{c} \text{slope of the plot} \\ \text{between Blue \& Green} \end{array} \right) \times \lambda_R \lambda_G$$

$$h_3 = \frac{e}{c} \times \left(\begin{array}{c} \text{slope of the plot} \\ \text{between Blue \& Red} \end{array} \right) \times \lambda_R \lambda_G \qquad h_{mean} = \frac{h_1 + h_2 + h_3}{3}$$

Graph :

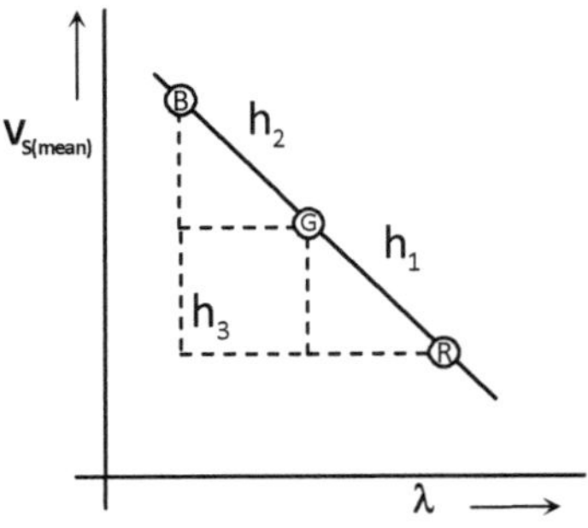

Fig: 5.2

Result :

Planck's constant determined using a Photocell is, h = ... Js

Theory :

- Planck's constant (h), a physical constant, was introduced by a German Physicist named Max Planck in 1900. Significance of Planck's constant is that 'quanta' (small packets of energy) can be determined by frequency of radiation and Planck's constant. It describes the behavior of particle and waves at atomic level as well as the particle nature of light.
- Principle of photoelectric effect is the core of this experiment.
- Photoelectric effect is the emission of electrons or other free carriers when light hits a material. Electrons such emitted are called as photoelectrons. Following equation governs the effect:

$$E = K.E + \phi$$

where, E is the total energy of the incident radiation, K.E. is the kinetic energy of the emitted electrons and ϕ is the work function of the cathode material.

- Stopping potential: It is the negative anode potential at which the photocurrent becomes zero (i.e. no photo electrons can reach the anode).

Precautions :

1. Source should be moved towards the photocell in a straight line to make the current zero.
2. Check photocell window is properly closed by cap; while adjusting the zero reading of the ammeter.
3. Replace the filter carefully.
4. Switch off the light source after the experiment.

Check Your Grasp :

1. *What is the standard value of Planck's constant?*
2. *What does 'h', the Planck's constant, represent?*
3. *What is the advantage of this method of finding value of h?*
4. *What is the relation between K.E. of photoelectrons and frequency of the incident radiation?*
5. *How intensity of the incident light would affect the experiment?*

 **Calculations :**

DETERMINATION OF WAVELENGTH OF LASER LIGHT

Aim : To determine the wavelength of Laser light using plane diffraction grating.

Apparatus : Laser source, Optical bench, Diffraction grating, Calibrated screen, Meter scale, etc.

Figure :

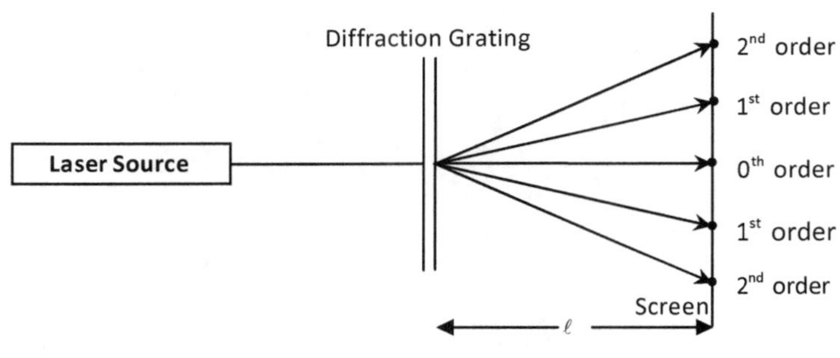

Fig. 6.1

Procedure :

1. Mount the given Laser source and diffraction grating on the optical bench and level the laser source in such a way that the laser beam strikes the grating **normally,** as shown in fig. 6.1.

2. Keep the suitable distance between the laser source and diffraction grating.

3. Adjust the distance between screen and the grating in such a way that it is greater than or equal to 1 meter.

4. Observe the diffraction pattern produced on the screen and measure the distance of each point from the centre of pattern on both the sides.

5. Tabulate the readings in the observation table.

68

 **Observations :**

- Distance between screen and the grating, ℓ = ……………….. cm
- Number of lines per inch of the grating, N = ……………….
- Therefore, grating element, e = (a + b) = $\dfrac{2.54}{N}$ cm = ……………….. cm

 **Observation Table :**

Order of diffraction (n)	Distance from the centre (x) (cm)			$\theta = \dfrac{180x}{\pi\ell}$ degree	sin θ	$\lambda = \dfrac{(a + b)\sin\theta}{n}$ (cm)	λ_{mean} (cm)
	L.H.S	R.H.S	Mean				
1							
2							
3							
4							
5							

 **Formulae :**

(a) Angle of diffraction, $\theta = \dfrac{180x}{\pi\ell}$ degree

(b) Wavelength of laser light, $\lambda = \dfrac{(a + b)\sin\theta}{n}$ Å

 **Result :**

Wavelength of given LASER light = ……………….. Å

 **Theory :**

- **Diffraction:** Bending of light at the sharp edges or the corners of an obstacle is called diffraction.
- Size of the obstacle should be comparable to the wavelength of light used.
- There are two types of diffraction as (a) Fresnel and (b) Fraunhofer diffraction

- **Diffraction grating:** It is a plane glass plate on which number of equidistant, parallel straight lines are etched. These lines divide the glass plate into alternate opaque and transparent regions. Ruled area becomes opaque whereas the region between two such lines remains transparent.

- The replica of original grating is used in the laboratory as the original gratings are very costly.

- A diffraction pattern is observed when light is incident on a grating. The condition for maxima in such a diffraction pattern is

$$(a + b) \sin \theta = n\lambda \qquad \qquad \text{... (1)}$$

and that for minima is

$$(a + b) \sin \theta = (2n + 1)\frac{\lambda}{2} \qquad \qquad \text{... (2)}$$

where $(a + b)$ is the grating element, given by $(a + b) = 2.54/N$ cm with N as number of lines per inch of the grating,

where a : width of transparent region of the grating, b: width of opaque region of the grating.

- LASER stands for Light Amplification by Stimulated Emission of Radiation.

- A laser beam has high directionality, very small divergence, greater intensity, high coherence as its characteristic properties. It has applications in the fields like engineering and technology, medical science, defense etc.

About the experiment:

- When the laser beam is focused on the grating, a diffraction pattern is produced on the screen. From the following figure, it can be seen that the angle of diffraction θ for m^{th} order fringe is given by:

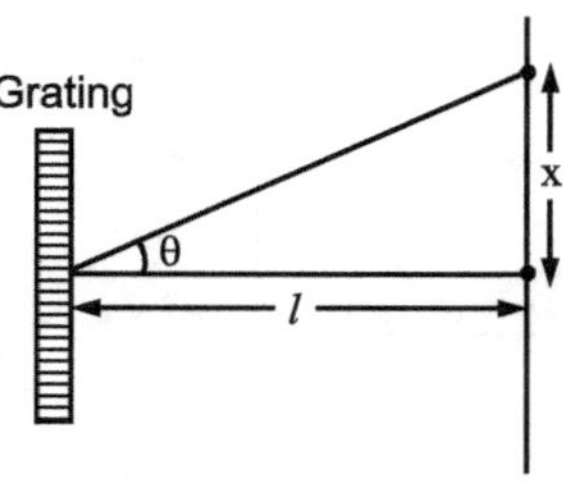

Fig. 6.2

From equation (1), if angle of diffraction; θ, order of diffraction; n, distances x and ℓ are known, wavelength of the laser source can be calculated using :

$$\lambda = \frac{(a + b) \sin \theta}{n} \qquad \qquad \text{... (3)}$$

Precautions :

1. Do not expose your naked eyes to the laser light.
2. If the angle of diffraction on L.H.S and R.H.S. are not the same, repeat the experiment.
3. Do not touch the ruled surface of the grating.
4. The distance between the grating and the screen should be at least 100 cm.

Check your grasp :

1. *What is diffraction?*
2. *How diffraction is different from refraction of light?*
3. *How diffraction is different from interference of light?*
4. *What are various types of diffraction?*
5. *Differentiate between the types of diffraction.*
6. *What is a diffraction grating?*
7. *Explain the construction of a diffraction grating.*
8. *How the laboratory diffraction grating is constructed?*
9. *If a grating of 500 lines per inch is used, in the present experiment, what will be the maximum possible order of the spectra with it?*
10. *What is LASER? Which are the characteristics and applications of LASER?*

Calculations :

STUDY OF I-V CHARACTERISTICS OF A SOLAR CELL

Aim : To study I-V characteristics of a given solar cell.

Apparatus : Solar cell, Variable load, Milliammeter, Millivoltmeter, Source of light (6W bulb) etc.

Figure :

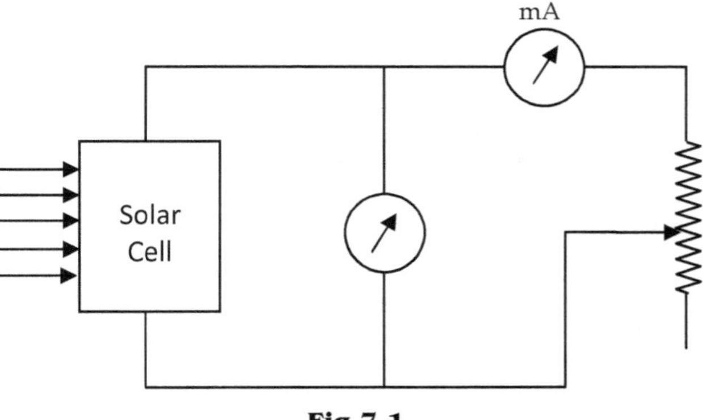

Fig.7.1

Procedure :

1. Connect the circuit as per circuit diagram.

2. Make the load resistance zero and note down the corresponding current in milliameter. This is short circuit current ($I_{Sc.}$)

3. Make the load resistance infinite and note down the voltmeter reading. This is open circuit voltage (V_{oc}).

4. Vary the load resistance from 0 Ω to 30 Ω and measure the corresponding current and voltage.

5. Plot a graph of current (I) vs voltage (V) which shows the IV characteristics of the solar cell.

73

6. Draw tangent to the plot at I_{sc} on the y axis and V_{oc} on the X axis

7. Find the values of I_m, V_m, I_{sc} and V_{oc} from the graph. Calculate the fill factor and efficiency of the solar cell using the given formulae.

 **Observations :**

Intensity of Incident light, I_L = ...candela

 **Observation Table :**

Obs No.	Resistance R (Ω)	Voltage V (mV)	Current I (mA)	Power (P = V×I) watt
1	0			
2	5			
3	10			
4	15			
5	20			
6	25			
7	30			

 **Formulae :**

Fill factor

$$\text{Fill Factor} = \frac{I_m \, V_m}{I_{SC} \, V_{OC}}$$

Efficiency

$$\eta = \frac{V_{oc} I_{sc} FF}{P_{in}}$$

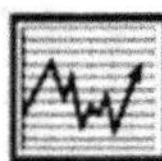 **Graph :**

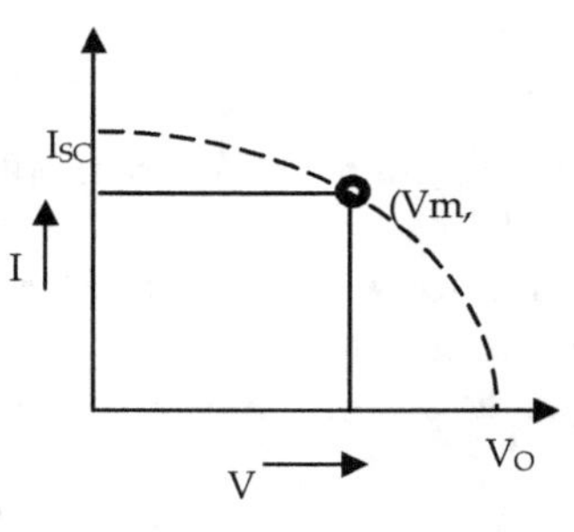

Graph – 1

Fig. 7.2

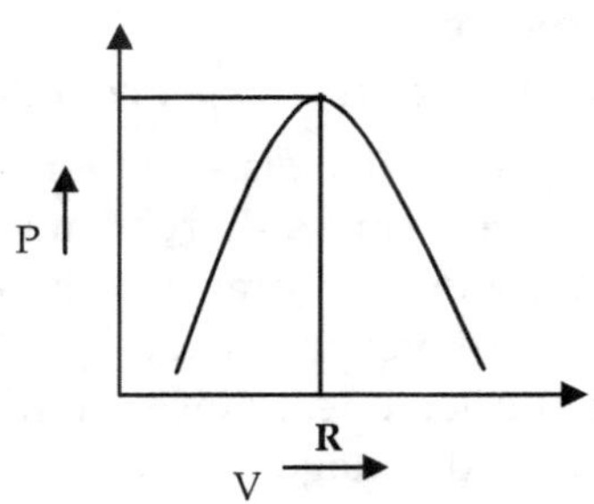

Graph– II

Fig. 7.3

Result :

- I-V characteristics of a given Solar cell are plotted as shown by curves.

- Fill factor for the solar cell is = ...

- Efficiency of the given solar cell is = ...

Theory :

- Solar cell is a P-N junction diode which converts solar energy into electrical energy. It is a sandwich of n-type and p-type semiconductor. It generates electricity by using sunlight to generate electrons and holes across the junction between the two semiconductors. The electron-hole pair will be generated when the incident photon has energy greater than the band gap of absorber material. The generated electron hole pair is separated by electric field created by p-n junction.

- Common materials used for solar cell are Silicon, Germanium, Gallium arsenide (GaAs), Indium Arsenide (InAs) and Cadmium arsenide (CdAs).

- *Solar Cell I-V Characteristics Curves* are basically a graphical representation of the operation of a solar cell.

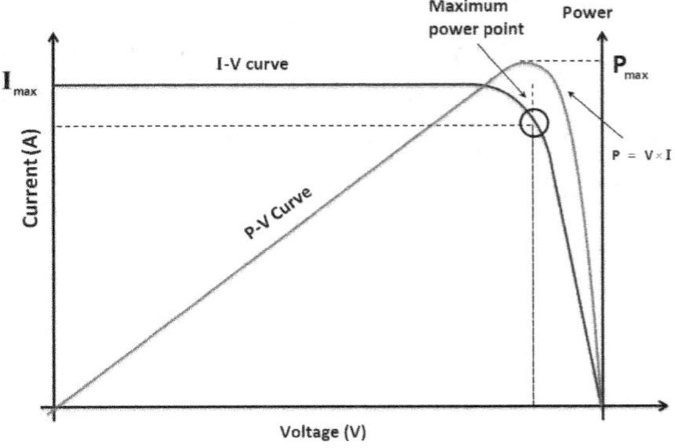

Fig. 7.4

- **Open circuit Voltage (Voc)** : This is the maximum voltage available from the solar cell and this is corresponds to zero current. The open circuit voltage corresponds to the amount of forward bias on solar cell due to the bias of the solar cell junction with light generated current.

$$V_{oc} = \frac{nKT}{q} \ln\left(\frac{I_L}{I_0} + 1\right)$$

where, n: ideality factor, K: Boltzmann constant, T: Temperature in Kelvin, q: Charge, I_0: Dark saturation current, I_L: Light generated current.

- **Short circuit current (I_{sc}):** This is the maximum current through the solar cell when the voltage across the solar cell is zero (when the solar cell is short circuited). The short circuit current is due to the generation and collection of charge carriers. The short circuit current depends on the area of solar cell, shunt and series resistance, incident light intensity, spectrum of incident light, optical losses etc.

$$J_{sc} = qG(Ln + Lp)$$

where, J_{sc}: Short circuit current density, G: Generation rate, L_n: Electron diffusion length, L_p: Hole diffusion length.

- **Fill Factor (FF):** This factor determines the maximum power from solar cell. It is the ratio of maximum power from the solar cell to the product of V_{oc} & I_{sc}.

$$FF = \frac{P_{max}}{V_{oc}I_{sc}}$$

$$P_{max} = V_m I_m$$

Actually, FF measures the squareness of the solar cell. FF depends on the parasitic resistances (shunt and series).

- **Efficiency (η):** It is the ratio of energy output from the solar cell to the input energy of the sun. The efficiency of solar cell depends on the intensity and spectrum of incident light and temperature of the solar cell.

$$\eta = \frac{V_{oc}I_{sc}FF}{P_{in}}$$

where, P_{in}: Input power

Precautions

1. Light from the source should fall normally on the solar cell.
2. The solar cell should be exposed to sun light before using it for the experiment.
3. Switch off the light source after the experiment.
4. Select the proper range of voltmeter and ammeter.
5. Short circuit while taking down reading of I_{sc} can lead to damage the cell permanently. Thus, longer duration of short circuit condition should be avoided.

 **Check Your Grasp :**

1. *What is the principle of solar cell and what is the difference between solar cell and photodiode?*
2. *What are the types of semiconductor materials used for solar cell?*
3. *What are the factors on which efficiency of solar cell depends?*
4. *What is dark current?*
5. *What do you mean by fill factor?*
6. *What is wavelength range of solar spectrum?*
7. *State applications of solar cell in our daily life.*

 **Calculations :**

III – 1

/ /

Experiment performed on

THEORETICAL STUDY OF CARNOT'S CYCLE

Aim : To interpret Carnot's cycle by drawing graphs of isothermal and adiabatic curves (P-V diagram).

Apparatus : Graph papers, Scale, Pencil, Calculator, etc.

Procedure :

1. Consider a particular system of reservoir set at initial temperature T_1.
2. Set the values of volume V_1 and pressure P_1 at the same temperature T_1.
3. The total process of Carnot cycle consists of four steps named sequentially as isothermal expansion, adiabatic expansion, isothermal compression and adiabatic compression.
4. In step-1 set the values of volume and pressure at constant temperature T_1 i.e. change V_1, P_1 to V_2, P_2 for same T_1. This is an isothermal expansion process.
5. In step-2 set the values of V_2, P_2 and T_1 to V_3, P_3 and T_2 respectively. This is an adiabatic expansion.
6. In the process of isothermal compression (step-3), the values of volume and pressure are set from V_3, P_3 to V_4 and P_4 at constant temperature T_2.
7. In the last step-4 i.e. in an adiabatic compression the volume, pressure and temperature are again set to initial values i.e. from V_4, P_4, T_2 to V_1, P_1 and T_1.
8. Collect the data from all above four steps and plot a graph of pressure against volume.
9. Indicate the temperature variations at each vertex of the graph in terms of co-ordinates (P, V and T).
10. Considering the trend of variation of all three parameters and the shape of the graph, draw your conclusions.
11. Compare your conclusions with theory.

Observation Table :

Obs No.	Steps	Processes	Stages	Pressure	Volume	Temperature
1.	I	Isothermal expansion	Initial	$P_1 =$	$V_1 =$	$T_1 =$
			Final	$P_2 =$	$V_2 =$	$T_1 =$
2.	II	Adiabatic expansion	Initial	$P_2 =$	$V_2 =$	$T_1 =$
			Final	$P_3 =$	$V_3 =$	$T_2 =$
3.	III	Isothermal compression	Initial	$P_3 =$	$V_3 =$	$T_2 =$
			Final	$P_4 =$	$V_4 =$	$T_2 =$
4.	IV	Adiabatic compression	Initial	$P_4 =$	$V_4 =$	$T_2 =$
			Final	$P_1 =$	$V_1 =$	$T_1 =$

Formulae :

- The heat received by air in isothermal expansion : $Q_1 = mART_1 \log(V_2/V_1)$ where, m is the mass of working substance (water), $A = 1/J$ is a constant, J is mechanical equivalent of heat and R is the gas constant.
- The heat rejected by air in isothermal expansion : $Q_2 = mART_2 \log(V_3/V_4)$
- Thermal efficiency :

$$\eta = \frac{\text{heat output}}{\text{heat input}} = \frac{Q_1 - Q_2}{Q_1} = \frac{T_1 - T_2}{T_1} \quad \therefore \quad h = 1 - \frac{T_2}{T_1}$$

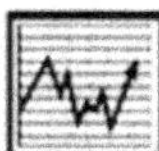

Graph :

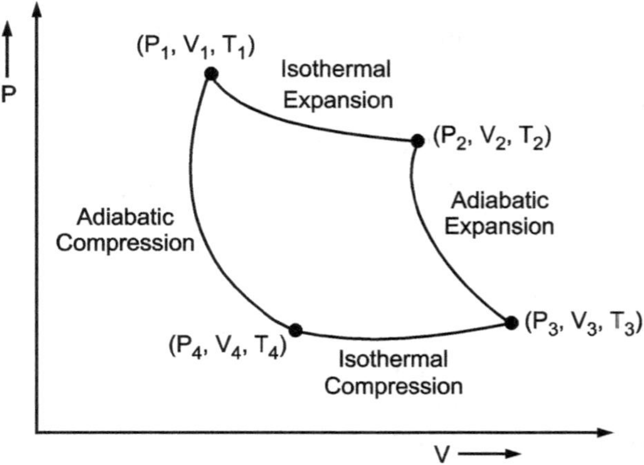

Fig. 1.1

Result :

Efficiency of the Carnot's engine calculated theoretically is = ...

Theory :

- It is well known that steam engines convert heat into mechanical work and its performance is measured in terms of its efficiency. This concept was first introduced by Carnot and is termed as Carnot cycle.

- In 1824, Carnot cycle represents a highly idealized process that is engaged in the continuous conversion of heat into work in a cyclic sequence of absorption of heat and its subsequent conversion to mechanical work. However, whereas a conventional steam engine employs steam as both the source of energy and the medium of its mechanical conversion, the Carnot cycle is based on a sealed container of ideal gas connected to a piston that captures the work performed. This may be either expansion or compression, depending on whether the container is placed in a hot or a cold reservoir, a cyclic sequence enabling the continuous nature of the conversion process. The Carnot's theorem states that between the two same temperature limits, a reversible engine has maximum efficiency i.e. no heat engine can work more efficiently than Carnot's engine working between the same temperature limits.

- The Carnot cycle is essentially comprised of the following steps carried out reversibly. The sealed container of one mole of ideal gas at volume V_1 is placed in a heat reservoir at a (higher) temperature T_1 and the gas allowed expanding isothermally to volume V_2, the container is removed from the reservoir, jacketed to prevent heat exchange, and the gas expanded adiabatically to temperature T_2 and volume V_3. The jacket is removed, and the cylinder is placed in a reservoir at a (lower) temperature T_2, and the gas compressed isothermally to volume V_4 the cylinder is again jacketed to prevent heat exchange, and the gas compressed adiabatically to the original volume V_1 and temperature T_1.

- Thus, apparently, the system has not only converted heat into work but also returned to the original state, ready to begin a new cycle. Note that in the two expansion steps, heat is both absorbed and mechanically expanded, into work performed upon the surroundings; conversely, in the compression steps work is performed upon the system which thus acquires heat. The splitting of the expansion and compression stages into isothermal and adiabatic steps is arbitrary, but apparently enables the calculation of the overall work performed in terms of its efficiency. From the formula of the efficiency mentioned above, it is clear that it depends only upon the temperature of source and sink and is independent of the nature of the working substance. The value of efficiency will be maximum if the value of T_2 is zero or value of T_1 is infinite. As this is not possible to achieve practically, hence the efficiency cannot be attained to 10%.

- The P-V diagram is normally displayed in a symmetrical manner implying that the opposite sides of the 'pseudo-parallelogram' are equal.

Precautions :

1. Select the working substance such that its standard minimum and maximum temperatures are well known viz water.
2. The initial values of P, V and T are to be chosen critically so that can be easily plotted on the graph paper.
3. The variation in the values of P, V and T must be selected in such a way that the P-V diagram on the graph paper shows vertex which clearly indicates the standard Carnot's cycle.
4. Indicate each line on the graph by their respective individual processes to avoid confusion regarding stages in the cycle.

Check Your Grasp :

1. *State Carnot's theorem.*
2. *What do you mean by (i) Heat sink, (ii) Source and (iii) Working substance?*
3. *What is the role of working substance in Carnot's cycle? How it affects efficiency?*
4. *What is maximum ideal efficiency of Carnot's cycle?*
5. *State applications of Carnot's cycle.*
6. *What do you mean by (i) isothermal expansion/compression, (ii) adiabatic expansion/compression?*

Calculations :

❖ ❖ ❖

STUDY OF TEMPERATURE COEFFICIENT OF THERMISTOR

Aim : To study temperature characteristics of Thermistor and to determine temperature coefficient of resistance.

Apparatus : Thermistor, multimeter, thermometer, heater, test tube, paraffin oil bath etc.

Figure :

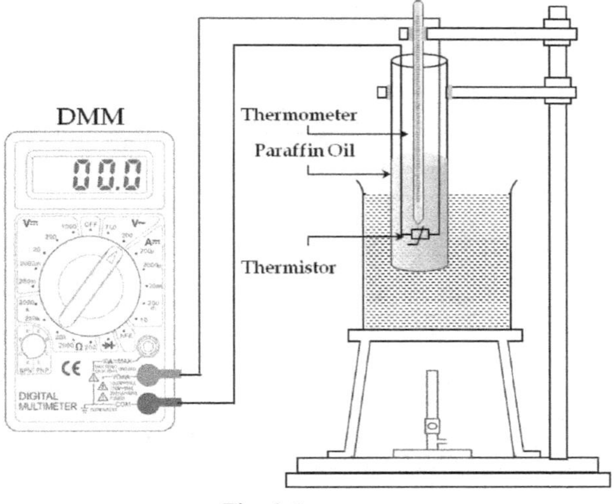

Fig. 2.1

Procedure :

1. Connect the circuit as shown in the figure and heat the paraffin bath to 90 ºC temperature.

2. Note down the resistance of the thermistor at temperature of 90 °C with the help of digital multimeter.

3. Record the successive readings of thermistor resistance with corresponding decrease in temperature in the steps of 5°C.

4. Plot a graph of ℓnR_T vs $(1/T)$ to calculate material specific constant (β) and calculate temperature coefficient of resistance of thermistors using the given formula.

 **Observations :**

- Room temperature, T_1 = ...°C.
- Resistance at room temperature, R_1 = ...kΩ.

 **Observation Table :**

Obs No	Temperature ($^\circ$C)	Temperature (K)	Resistance (R_T) (Ω)	ℓn R_T	$\ell nR_T - \ell nR_1$	$\dfrac{1}{T}$ (K^{-1})	$\dfrac{1}{T} - \dfrac{1}{T_1}$ (K^{-1})	$\beta = \dfrac{(\ell nR_T - \ell nR_1)}{\left(\dfrac{1}{T} - \dfrac{1}{T_1}\right)}$	$\alpha = -\dfrac{\beta}{T^2}$ (K^{-1})
1	90								
2	85								
3	80								
4	...								
5	...								
6	...								
7	...								
8									
9									
10									
							Mean		

Formulae :

Material specific constant (β):

By calculation: $\beta = \dfrac{\left(\ell n\, R_T - \ell n\, R_1\right)}{\left(\dfrac{1}{T} - \dfrac{1}{T_1}\right)}$

where, R_T = resistance at T °C, T_1 = reference temperature (room temperature), R_1 = resistance of thermistor at room temperature.

By calculation: β = slope of $\left(\ell n\, R_T - \ell n\, R_1\right)$ vs $\left(\dfrac{1}{T} - \dfrac{1}{T_1}\right)$

The temperature coefficient of thermistor (α):

$$\alpha = -\beta/T^2 \qquad (K^{-1})$$

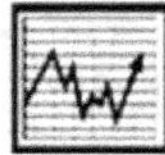

Graph :

Plot a graph of $\ell n R_T$ vs $1/T$ to calculate material specific constant.

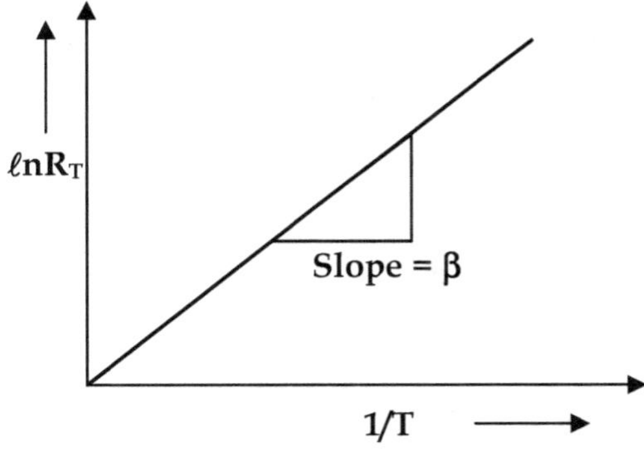

Fig. 2.2

Results :

- The material specific constant, β = ...
- The temperature coefficient of resistance of thermistor, α = .../°C

Theory :

- **Thermistor**: It is the thermally sensitive resistor. It is mainly used as temperature sensor, current limiter, self-resetting over current protector and self-regulating heating element.
- There are many types of thermistors, mainly they are classified as positive temperature coefficient of resistance (PTC) and negative temperature coefficient of resistance (NTC).
- **PTC**: Thermistors in which resistance of the material increases with the increase in temperature, are said to have PTC. Here, slope of the characteristic curve is positive. e.g. all metals have PTC.
- **NTC**: Thermistors in which resistance of the material decreases with the increase in temperature, are said to have NTC. Here, slope of the characteristics curve is negative. e.g. semiconducting materials have NTC.
- **Thermistor characteristics**: Thermistor characteristics is nonlinear in nature, which satisfies the relation

$$R = R_0 e^{\beta\left(\frac{1}{T} - \frac{1}{T_0}\right)} \qquad \qquad ...(1)$$

- β is the material specific constant lies between 3000 to 4400 K.
- Taking ln on both side of equation 1 we get,

$$\ln R = \ln R_0 + \beta\left(\frac{1}{T} - \frac{1}{T_1}\right) \qquad \qquad ...(2)$$

- In solving equation to for β we get β as

$$\beta = \frac{\ln R - \ln R_0}{\dfrac{1}{T} - \dfrac{1}{T_0}} \qquad \qquad ...(3)$$

- On differentiation with respect to T we get

$$\frac{1}{R}\frac{dR}{dT} = -\frac{\beta}{T^2}$$

but this is temperature coefficient of resistance that is α.

Precautions :

1. Thermometer must be placed in paraffin bath not in water bath.
2. Don't add water into water bath while taking cooling readings.

 **Check Your Grasp :**

1. *What do you mean by thermistor?*
2. *What is temperature coefficient of resistance?*
3. *Define NTC and PTC and give their examples.*
4. *Why resistance of semiconductor material decreases with the increase in temperature?*
5. *What are the applications of thermistor?*
6. *Why readings are always take during cooling cycle?*
7. *How to draw calibration curve for thermistor?*
8. *Explain the types of thermistor.*

 **Calculations :**

STUDY OF THERMOCOUPLE

Aim : To study a thermocouple and determination of inversion temperature.

Apparatus : Copper constantan thermocouple, Clamp stand, Water bath, Heating arrangement, Funnel with stand, Beaker, Ice, Thermometer, Potentiometer, Galvanometer, Resistance box, Standard cell, Key, Jockey, Voltmeter, Connecting wires, etc.

Figure :

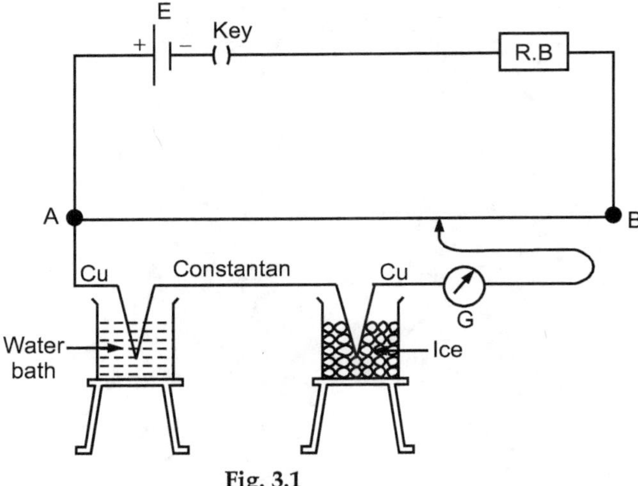

Fig. 3.1

Procedure :

1. Connect the circuit components as shown in Fig. 3.1.
2. Read all the precautions before starting the experiment.
3. Clamp the thermo-couple on a stand. Dip one junction in water bath and immerse the other junction in ice as shown in Fig. 3.1.

4. Heat the water bath till the water boils. Keep suitable resistance ($\sim$ 2000 Ω) in the resistance box. Close the key K and obtain the balance point.

5. Stop heating the water bath and allow the temperature to fall. Obtain the balance point at temperatures 90 °C, 85 °C, 75 °C and so on.

6. Plot a graph between temperature of the hot junction and thermo-electric e.m.f.

7. Find out the neutral temperature from the graph and then calculate inversion temperature using the given formula.

 **Observations :**

- Resistance of potentiometer wire, r = ... Ω
- E.M.F. of the standard cell, E = ... V
- Resistance taken out from resistance box, R = ... Ω
- Length of the potentiometer wire, L =.................. cm
- Potential gradient, k =.................. V/cm

 **Observation Table :**

Obs. No.	Temperature of the hot junction (°C)	Balancing length (ℓ) (cm)	Voltage across CB (E_L) (volt)	Thermo e.m.f $e = k \times \ell$ (volt)
1	90			
2	85			
3	80			
4	75			
5	70			
6	65			
7	60			
8	55			
9	50			
10	45			
11	40			
12	35			

 **Formulae :**

Temperature of inversion, $\theta_i = 2\theta_n - \theta_c$

where, θ_n: neutral temperature, θ_c: temperature of cold junction

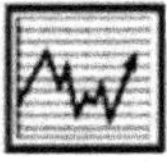 **Graph :**

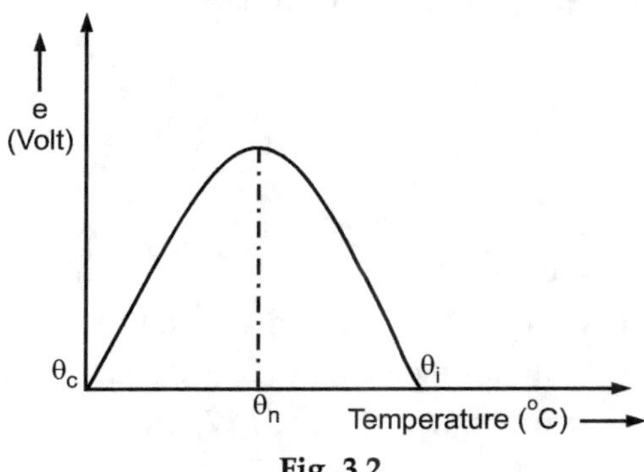

Fig. 3.2

 **Result :**

Inversion temperature, θ_i = ... ^{o}C

 **Theory :**

- The closed circuit formed by the wires of two different metals is known as thermocouple.
- The magnitude of thermo e.m.f depends upon the temperature difference between the two junctions and the two metals that form a thermocouple.
- Variation of thermo-e.m.f with temperature shows that the thermo-e.m.f increases with increase in the temperature of the hot junction, becomes maximum for a particular temperature θ_η and then decreases and changes its direction at inversion temperature, θ_i.
- **Neutral temperature :**
  - The temperature of the hot junction at which the thermo-e.m.f becomes maximum is called neutral temperature.
  - It is independent of temperature of the cold junction. It depends only upon the nature of the metals constituting the thermocouple.

- **Inversion temperature :**
  - The temperature at which the thermo-e.m.f becomes zero and then changes its direction (gets inverted) is called the temperature of inversion.
  - It depends upon temperature of the cold junction.
  - The neutral temperature and the inversion temperature are related by the relation $\theta_i = 2\theta_n - \theta_c$
- **About the experiment:**
  Consider the experimental arrangement as shown in the circuit diagram.
  Let E be the e.m.f of the standard cell, L: total length of the potentiometer wire, r : resistance of the potentiometer wire. Then the current through the wire is,

$$I = \frac{E}{R + r} \qquad \dots (1)$$

The potential difference across the wire is,

$$V = Ir = \frac{E}{R + r} \qquad \dots (2)$$

The potential gradient along the wire is,

$$k = \frac{V}{L} = \frac{Er}{R + r} \cdot \frac{\ell}{L} \qquad \dots (3)$$

The thermo e.m.f. (e) is the potential drop across the balancing length.

$$e = k \times \ell = \frac{Er}{R + r} \cdot \frac{\ell}{L} \qquad \dots (4)$$

This gives the thermo e.m.f. at a particular temperature. Plot of temperature versus thermo e.m.f. then gives the neutral and inversion temperature.

 **Precautions :**

1. The thermo-couple should be connected to the potentiometer such that the direction of the thermoelectric current in the potentiometer wire is same as that of current due to the standard cell used.
2. Since the thermo e.m.f. produced is usually small (in micro volt), poor contacts lead to errors. Hence terminals of the connecting wires must be clean and the connections should be tight.
3. The thermo-couple should be carefully checked before the experiment to confirm that the junctions are properly soldered.
4. As the e.m.f. of the standard cell has to remain constant during the experiment, it should be freshly charged before the experiment.
5. Resistance R should be so adjusted so that the balance point for the highest temperature of the water bath is obtained on the last wire.
6. The positive of the thermo-couple should be connected to the point where the positive of the cell is connected.

Check Your Grasp :

1. *What do you mean by (i) thermoelectricity, (ii) thermocouple, (iii) Seebeck effect, (iv) Peltier effect and (v) Thomson effect?*
2. *Define: (i) Neutral temperature, (ii) Temperature of Inversion, (iii) Thermoelectric power.*
3. *What is the relation between neutral temperature and temperature of inversion?*
4. *What should be the order of the potential gradient along the potentiometer wire so that the balancing length for the highest temperature lies on the last wire of the potentiometer?*
5. *Why a standard cell is used in the experiment?*
6. *How you will connect the copper constantan thermo-couple to the potentiometer?*

Calculations :

THERMAL CONDUCTIVITY BY LEE'S METHOD

Aim : To determine the coefficient of thermal conductivity of a bad conductor by Lee's method.

Apparatus : Lee's apparatus, Bad conductor in the form of circular disc, Vernier calliper, Micrometer screw gauge, Steam generator, Thermometers, Stop watch, etc

Figure :

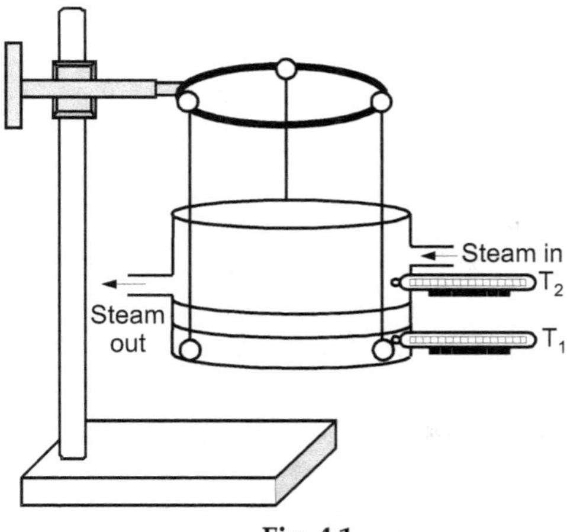

Fig. 4.1

Procedure :

1. Arrange the apparatus as shown in Fig. 4.1.
2. Insert the bad conductor sheet between the steam jacket and metal slab.
3. Ensure no air gap between the steam jacket, metal slab and the bad conductor.
4. Insert thermometers in steam jacket and metal slab.
5. Now, pass a steam through the steam jacket. Observe a change in the temperature of the steam jacket and metal slab.

6. Note the steady state temperatures of the steam jacket (θ_1') and metal slab (θ_2').

7. Now interchange the thermometers and wait till the thermometers show steady temperatures. Note down the steady state temperature of the steam jacket (θ_2'') and metal slab (θ_2'').

8. Remove the bad conductor safely.

9. Keep the steam jacket in direct contact with the metal slab till the temperature of the metal slab increases by 10 ºC.

10. Remove the steam jacket and note down the temperature of the metal slab (θ) after every 30 seconds during its cooling.

11. Plot a cooling curve of temperature (θ) versus time (t).

12. Find out the slope of the cooling curve at a temperature θ_2 and calculate thermal conductivity of the given bad conductor.

 **Observations :**

- Least count of vernier calliper, L.C. = ...cm
- Least count of micrometer Screw Gauge, L.C. = ...cm
- Mass of the metal slab, m =....................g.
- Specific heat of material of metal slab, s = ...cal/gºC
- Diameter of the bad conductor:
    (i) D_1 = ... cm (ii) D_2 =............... cm (iii) D_3 = ... cm
    Mean diameter, D = ... cm
- Radius of the bad conductor, r = D/2 = ... cm
- Thickness of the bad conductor,
    (i) d_1 =............... cm (ii) d_2 = ... cm (iii) d_3 =............... cm
    Mean thickness, d = ... cm

 **Observation Table :**

Part - I

Obs. No.	Temperature of	Steady State Temperature		Temperature (ºC)	Difference $(\theta_1 - \theta_2)$ (ºC)
		before interchanging (ºC)	after interchanging (ºC)		
1	Steam Jacket	θ_1' =	θ_1'' =	θ_1 =	
2	Metal Slab	θ_2' =	θ_2'' =	θ_2 =	

Part - II

Obs. No.	Time t (s)	Temp. θ (ºC)	Obs. No.	Time t (s)	Temp. θ (ºC)	Obs. No.	Time t (s)	Temp. θ (ºC)

Formulae :

Coefficient of thermal conductivity, $k = \dfrac{m \cdot s}{\pi r^2} \dfrac{d}{(\theta_1 - \theta_2)} \left(\dfrac{d\theta}{dt}\right)_{\theta_2}$ cal/cm $\cdot$ s ºC

where, m : mass of the metal slab, s : specific heat of material of the bad conductor, r : mean radius, d : thickness of the bad conductor, θ_1 and θ_2 : temperature limits.

Graph :

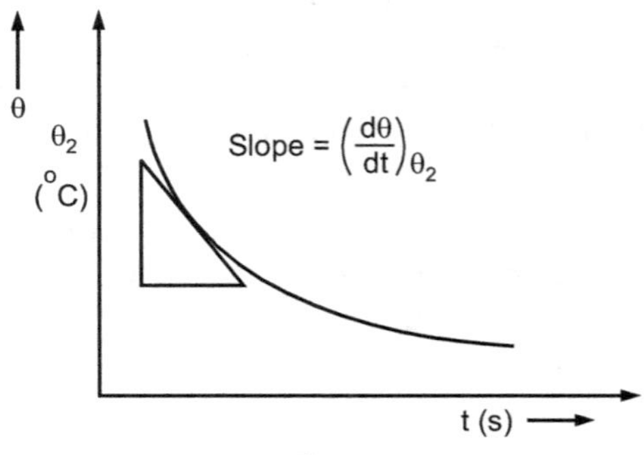

Fig. 4.2

Result :

Coefficient of thermal conductivity of the bad conductor, k =…………….cal/cm·s ºC

Theory :

- Heat is a form of energy which is when supplied to a body, is stored by itself in the form of heat or thermal energy, due to which temperature of the body increases. This heat energy may be supplied to all forms of matter i.e. solid, liquid or gases in two different ways as conversion and transfer of energy. The transfer of energy always takes place from a body at higher temperature to a body at lower temperature.
- The transfer of heat takes place in three different modes i.e. (i) Radiation, (ii) Convection and (iii) Conduction.
- The radiation is the process of transmission of heat energy between two bodies without any interconnecting medium, as in case of transfer of heat from a lamp.
- In convection, the heat transfer takes place from one part of the medium to another by actual migration of heated particles as in case of boiling water.
- Conduction is the mode of heat transfer in which heat is transferred from the region of higher temperature to the one at lower temperature without any motion of particles, as in case of heating up of a metallic rod.
- Thermal conductivity of a material is the physical property of matter which indicates the quantity of heat transferred through the material and is defined as *flow of heat in one second through unit cube of the material when its opposite faces are at a temperature difference of 1 ºC.*
- The value of thermal conductivity indicates whether the material is a good or bad conductor.
- There are various methods to determine the thermal conductivity of a material. A method developed by Lee is particularly used to measure thermal conductivity at lower temperatures i.e. below 100 ºC
- In first part of the experiment, steam is used to increase the temperature of a chamber upto a saturation limit. The heat is transferred from this chamber to a small metallic disc through a bad conductor like a card board. In the other part, the bad conductor is removed and the metallic disc is kept in direct contact with the steam chamber so that the heat is directly passed on to the disc. The temperature difference between these two steps indicates the effect of bad conductor which is attributed to its thermal conductivity.

Precautions :

1. Ensure that the bad conductor has uniform thickness.
2. Diameters of the bad conductor, steel jacket and metal slab should be same.
3. Steam should not leak out from the apparatus.
4. Do not touch the steam jacket or the metal slab directly.
5. Measure the thickness of the bad conductor accurately.
6. Both the thermometers should be identical. Handle them carefully.

Check your grasp :

1. *What is heat? What is the difference between heat and temperature?*
2. *What do you mean by thermal conductivity?*
3. *What is the working principle of Lee's method?*
4. *What are the different modes of heat transfer? Distinguish between them with suitable example.*
5. *What do you mean by steady state of temperature?*
6. *Why there is a steady state?*
7. *Why materials are good or bad conductors of heat?*

Calculations :

SPECIFIC HEAT OF GRAPHITE

Aim : To study specific heat of graphite.

Apparatus : Specific heat kit, Dimmerstat, AC Voltmeter (0–300 V), AC Ammeter (0-10 A), DC Millivoltmeter (0–30 mV), Chromel Alumel thermocouple, Galvanometer, Lamp and Scale arrangement.

Figure :

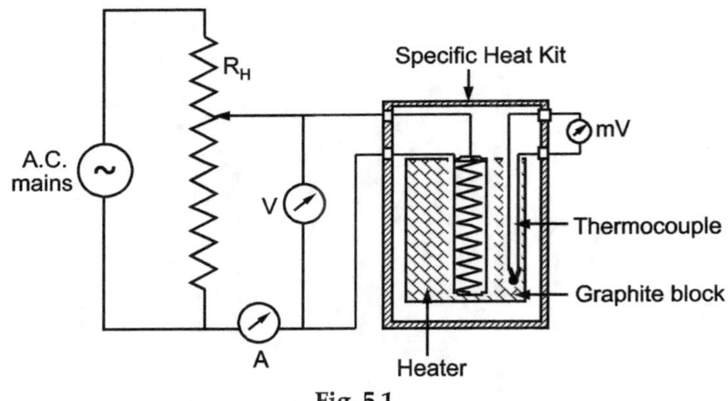

Fig. 5.1

Procedure :

1. Connect the specific heat kit and other circuit components as shown in fig.5.1.
2. Keep the P.D applied to the heater at a lower (~100 V) value untill an equilibrium temperature is obtained. Note the equilibrium temperature, voltage, current through the circuit.
3. Increase heater voltage in steps of 20 V upto 200 V. Note down the equilibrium temperature, voltage, current in each case. Tabulate the readings in Table (1).

4. Switch off the heater once the equilibrium temperature for 200 V is obtained. Allow the block to cool down and note down its temperature after every three minutes, in Table (2). Plot a graph of temperature versus time (cooling curve).

5. Calculate the slope of the cooling curve at each equilibrium temperature and calculate the specific heat at these temperatures.

6. Plot a graph between the specific heat and the temperature.

 **Observation Table**

Table (1)

Obs. No.	Temperature (°C)	Voltage (V) (volt)	Current (I) (A)	Power (P) P = V × I (W)
1				
2				
3				
4				
5				

Table (2)

Obs. No.	Time (min)	Temperature (°C)	Obs. No.	Time (min)	Temperature (°C)
1			12		
2			13		
3			14		
4			15		
5			16		
6			17		
7			18		
8			19		
9			20		
10			21		
11			22		

Table (3)

Obs. No.	Power (W)	$\dfrac{d\theta}{dt}$ (°/sec)	Specific Heat (s) cal/g°C
1			
2			
3			
4			
5			

Formulae :

Specific heat of graphite is given by

$$s = \frac{VI}{m \times slope \times J} \quad cal/g°C$$

where, V: P.D applied to the heater, I: Current through the heater coil, m: mass of the graphite block, J : Joule's constant = 4.18 J/cal.

Graph :

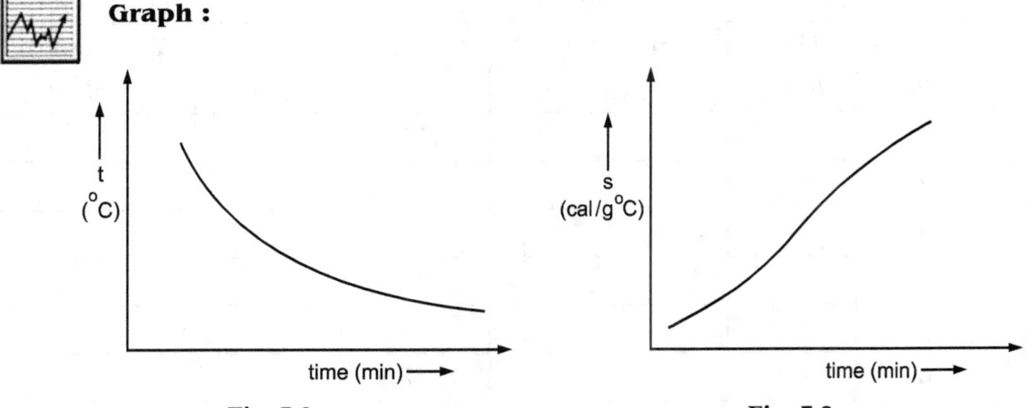

Result :

Specific heat of graphite, s = ... cal/g°C

 **Theory :**

- **Specific heat :** *It is the amount of heat required to raise the temperature of unit mass of a substance through 1ºC temperature.*

- At lower temperatures the heat capacity drops markedly, hence the measurement of specific heat of a substance at various temperatures is important. The variation of specific heat with temperature can be attributed to atomic vibrations in the crystals and outermost electrons of the atoms. Usually contribution from electrons is small relative to that of lattice vibrations.

- When a block of graphite of mass m, specific heat 's', is heated with an electrical heater the quantity of heat Q given to the system is expressed by

$$Q = ms\theta$$

where θ : change in temperature.

$$\therefore \qquad \frac{dQ}{dt} = ms\frac{dq}{dt} \qquad \qquad \text{... (1)}$$

- When the heater is turned off and the block is allowed to cool, the rate of change of temperature is given by $d\theta/dt$. This can be obtained from the rate of cooling given by slope of the cooling curve $(d\theta/dt)$ at equilibrium state temperature of the block. At this temperature, power of the heater is P = IV (I is the current through the heater and V is the P.D. across the heater coil, So, we can write,

$$\therefore \qquad \frac{dQ}{dt} = \frac{P}{J} = \frac{VI}{J} \qquad \qquad \text{... (2)}$$

where J is the electrical equivalent of heat and is proportional to the power P supplied to the heater.

Thus, if mass m of the block is known, the specific heat s at any temperature can be given by

$$\therefore \qquad s = \frac{VI}{m \times slope \times J} \qquad \text{c.g.s units} \qquad \text{... (3)}$$

- The specific heat obtained from the above relation is at a particular temperature. From these values, a graph of specific heat versus temperature can be plotted.

- **Specific heat kit :** It is a closed steel box lined with asbestos sheet, containing a block of graphite. The graphite block has a central hole drilled in the block to insert the heater. The heater has a handle to pull it out whenever necessary. One smaller hole drilled to the graphite block provides the space for the Chromel - Alumel thermocouple.

 **Precautions :**

1. Voltage applied to the heater should not exceed 200 volts to avoid its damage.
2. Temperature measurement using the thermocouple should be very accurate.
3. The term $d\theta/dt$ should be calculated very precisely.
4. Keep the graphite box lid open during its cooling.

 **Check Your Grasp :**

1. *Define specific heat.*
2. *How specific heat differs from latent heat?*
3. *What are the factors on which specific heat of a material depends?*
4. *Is graphite, a purely carbon based material?*
5. *What is thermal capacity? How it is related to specific heat of a material?*
6. *What are the various methods to determine specific heat of a material?*

 **Calculations :**

STUDY OF SOLAR CONSTANT

Aim : To determine the solar constant

Apparatus : 2 L Bottle, Water, Thermometer, Stop watch, Black ink etc.

Figure :

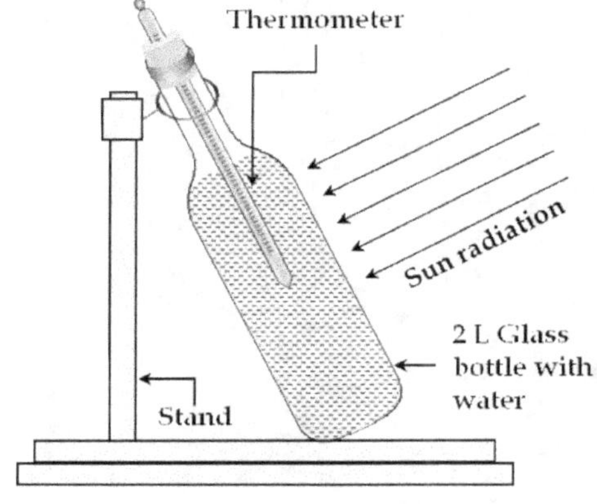

Fig. 6.1

Procedure :

1. Fill the glass bottle with 150 ml of water. Add a few drops of the black ink to make colour of water black, and put the thermometer through the hole in the cork.
2. Make sure the cork seal is as tight.
3. Put the collector bottle in shade to record its steady state temperature and note down its steady state temperature.
4. Move the bottle collector into sunlight, place it in such a way that solar radiation incident on it should be normal to its surface area.

5. Allow the collector in sunlight until its temperature rise by 4 to 5 °C and note down the time to increase this much temperature.

6. Measure the exposure area of the bottle collector.

 **Observations :**

- Volume of water, V = ...liters
- Mass of water used, M = ...Kg.
- Exposed surface area, A = ... m²
- The specific heat of water, s = 4186 J/(kg °C).
- The correction factor for Low humidity, completely clear sky, Fc = 1.4.

 **Observation Table :**

Obs.No.	Initial temperature (°C)	Final temperature (°C)	ΔT (°C)	Elapsed Time t (S)	ΔT/t (°C/s)
1					
2					
3					
				Mean	

 **Formulae :**

The energy absorbed by water per second is given by,

$$Q = s \times M \times \left(\frac{\Delta T}{t} \right)_{mean} \quad J/S$$

where, s: specific heat of water and M mass of water taken.

Average energy collected per unit of surface area is

$$E_{ave.} = Q/A \quad J/s.m^2$$

where, A is exposure area.

Solar constant = $E_{ave.} \times 2 \times Fc.$ W/m².

The multiply factor 2 is to correct uncorrected solar irradiation for the glass.

Result :

The solar constant is = ...W/m^2.

Theory :

- **Solar energy:** The energy radiated by the sun in the form of electromagnetic radiation is called as solar energy. It is radiated in terms of light and thermal energy.
- **Radiation:** The process of transfer of heat from one place to another along a straight line without affecting an intervening medium is called radiation.
- **Thermal radiation:** The energy emitted by a body in the form of radiation by virtue of its temperature is called thermal radiation of energy.
- **Solar constant:** It is the rate at which energy reaches the earth's surface from the sun, usually taken to be 1,388 W/m^2.
- **Solar collectors:** Solar collector is a device used to collect solar radiation and transfer it to fluid in its contact. These are generally classified as concentrating and non-concentrating collectors. In concentrating type of collector, solar radiation is collected at the focal point of the collector to get maximum concentrated radiation.
- **Exposure area:** Area over which the solar radiation incident is called exposure area.

Precautions :

1. Ensure that the bottle collector is air tight.
2. Bottle should be inclined in such a way that solar radiation must incident normally on its surface.

Check Your Grasp :

1. *What do you mean by solar radiation?*
2. *What is exposure area?*
3. *What is radiation?*
4. *Define solar constant.*
5. *What is thermal radiation?*
6. *What is greenhouse effect?*
7. *List the instruments used to measure different radiation parameters.*

 **Calculations :**

DETERMINATION OF CALORIFIC VALUES OF DIFFERENT FUELS

Aim : To calculate calorific values of various fuels.

Apparatus : Thermometer, Water, Stirrer, Calorimeter, Cow dung& other fuels

Figure :

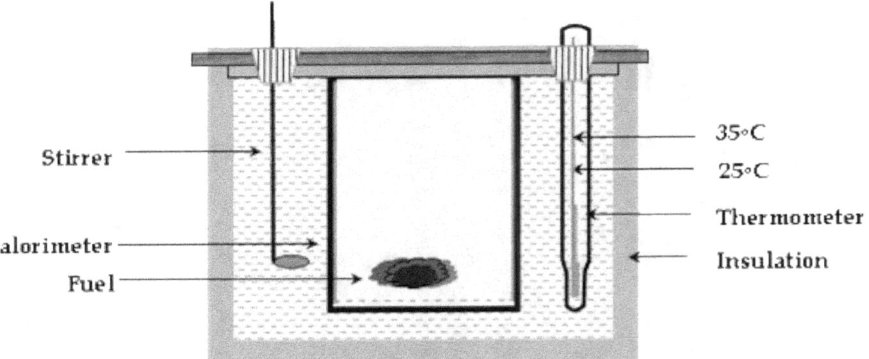

Fig. : 7.1

Procedure :

1. Measure the room temperature and note down.
2. Take a weight of cow dung and note down.
3. Add 1000 ml water in the Bomb calorimeter and stirred by using stirrer.
4. Ignite the cow dung by using candle.
5. Stirred the water regularly for making constant temperature in calorimeter. A constant rate of stirring is used.
6. Measure the temperature of the water at every 5 minutes

---

7. Continue this process up to temperature increases 20°C from the room temperature.
8. Open the cap of calorimeter and remove the whole water from the calorimeter.
9. Open the cap of calorimeter and remove the unburned cow dung.
10. Measure the weight of unburned cow dung and note down.
11. Determine the fuel value by using the formula.
12. Repeat the entire procedure for two more fuels.

 **Observations :**

Part 1

*   Specific heat of material of calorimeter & stirrer (s_1) = ... cal / g°C
*   Specific heat of water (s_2) =   1 cal/g°C
*   Mass of and stirrer (m_1) = ... gm
*   Mass of water taken (m_2) = 1000 ml = 1084 g

Part 2

*   Mass of cow dung taken = ... gm
*   Initial temperature of water (θ_1) = ... °c
*   Initial temperature of water (θ_2) = θ_1 + 18 °c = ...
*   Mass of unburned fuel (W)= .... gm
*   Mass of burned fuel = ... gm

 **Formulae :**

(a) Amount of heat energy generated: $H = m_1 \times s_1 \times (\theta_2 - \theta_1) + m_1 \times s_2 \times (\theta_2 - \theta$ (cal)
(b) Fuel value = H/W (cal/gram)

 **Result :**

*   **Amount of heat generated H =..................cal**

*   **Fuel value of cow dung =...................cal/gram**

Theory :

- Calorimetry is the science of measuring changes in state variables of a body for the purpose of deriving the heat transfer associated with changes of its state. A bomb calorimeter is a type of constant-volume calorimeter used in measuring the heat of combustion of a particular reaction. Bomb calorimeters have to withstand the large pressure within the calorimeter as the reaction is being measured. The principle of operation is to saturate the material with oxygen, within a sealed container (the bomb) and ignite. During the rapid combustion carbon molecules are converted to carbon dioxide, hydrogen to water and nitrogen to gaseous nitrogen. The recorded enthalpy change is thus a sum of all bonds broken and bonds made converting the organic solid into simple gaseous molecules. The measured enthalpy change for this type of analysis is often referred to as the enthalpy of formation. Note that, in fact, this type of calorimeter operates at constant volume. As such, the energy measured is internal energy change (ΔU) and not enthalpy change (ΔH).

- Bomb calorimeter is an apparatus primarily used for measuring heats of combustion. The reaction takes place in a closed space known as the calorimeter proper, in controlled thermal contact with its surroundings, the jacket, at constant temperature. This set, together with devices for temperature measurement, heating, cooling, and stirring comprise the calorimeter. The calorimeter proper is usually a metal can with a tightly fitting lid containing water, stirred continually, in which the bomb itself is situated. It consists of a sealed heavy-walled container in which the reactants are allowed to react, under constant volume conditions, following the ignition of the combustible matter in an oxygen atmosphere.

Precautions :

1. Bomb calorimeter should be placed above the ground level.
2. Cow dung should be dry and weighable.
3. Carefully add the water in the calorimeter.
4. Carefully ignite the cow dung in the calorimeter.
5. Clean the calorimeter after the experiment.
6. Apparatus should be air tight.

Check Your Grasp :

1. *Define: Renewable energy and Non-renewable energy*
2. *What are the various methods to determine calorific value of a fuel?*

3. *What are the factors on which fuel value of a fuel depends?*
4. *What do you mean by calorimetry?*
5. *Why this instrument is called Bomb calorimeter?*
6. *What are the part of calorimeter and its function?*

 **Calculations :**

IV – 1

/ /

Experiment performed on

CHARGING AND DISCHARGING OF A CAPACITOR

Aim : To study charging and discharging of a capacitor and to determine RC time constant of a given circuit.

Apparatus : Capacitor, Resistor, Battery, Regulated power supply, DC Voltmeter, Four-way key etc

Figure :

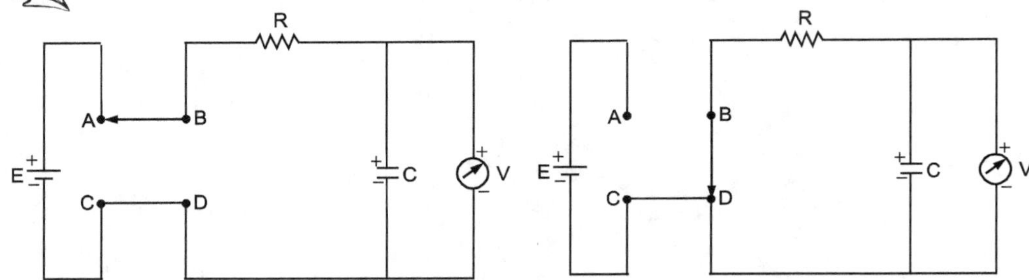

Fig. 1.1 : For charging   Fig. 1.2 : For discharging

Procedure :

Part I: Charging

1. Connect the circuit as shown in Fig. 1.1.
2. Close the switches AB and CD so as to charge the capacitor.
3. Note down the voltage across the capacitor after every 5 seconds. Continue the readings till the voltage across the capacitor become constant.
4. Plot a graph of voltage across the capacitor against time. From the graph, note down the maximum voltage across the capacitor.
5. Determine the RC time constant from the graph as well as by calculations.

Part II: Discharging

1. Connect the circuit as shown in Fig. 1.2.
2. Close the switches BD and CD so as to discharge the capacitor.
3. Note down the voltage across the capacitor after every 5 seconds. Continue the readings till the voltage across the capacitor become zero.
4. Plot a graph of voltage across the capacitor against time. From the graph, note down the maximum voltage across the capacitor.
5. Determine the RC time constant theoretically and also from the graph.

Observations :

- R = ... Ω

- C = ... μF

Observation Table :

For charging of capacitor

Obs. No.	Time (T) sec	Voltage (V) volt

For discharging of capacitor

Obs. No.	Time (T) sec	Voltage (V) volt

 **Formulae :**

- RC time constant (Theoretical) $T_{RC} = (R \times C)$ sec

By graph:

- For charging, $T_{RC} = 0.632 \times V_{max(C)}$ sec
- For discharging, $T_{RC} = 0.368 \times V_{max(D)}$ sec

  where, $V_{max(C)}$: peak voltage across the capacitor while charging

  $V_{max(D)}$: peak voltage across the capacitor while discharging.

 **Graph :**

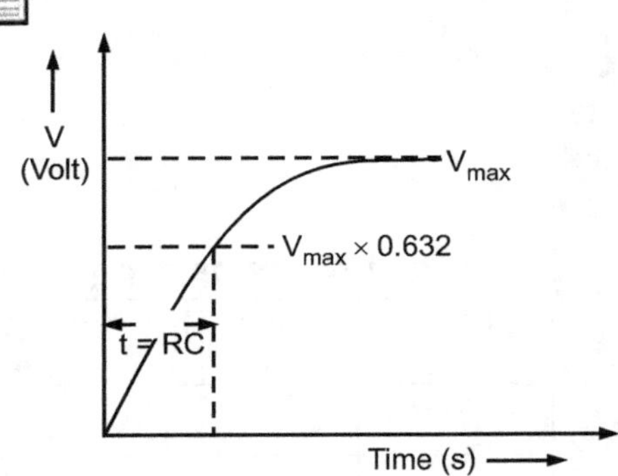

Fig. 1.3 For charging

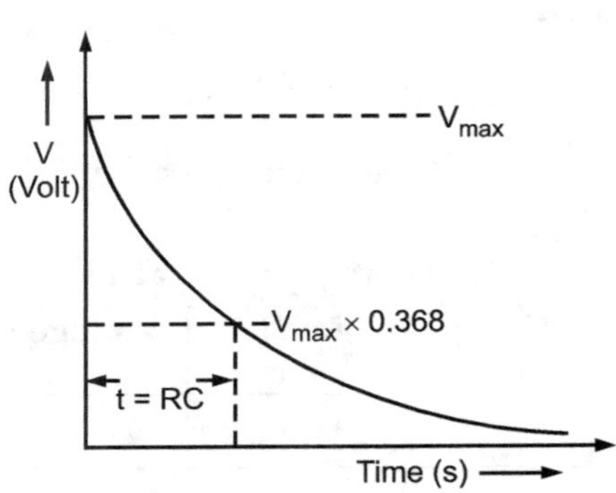

Fig. 1.4 For discharging

 **Result :**

- RC time constant (Theoretical) =................... sec
- RC time constant (by graph):
    - For charging the capacitor =.................... sec
    - For discharging the capacitor =............... sec

 **Theory :**

- **Charging a capacitor :**
  When e.m.f is applied to a capacitor, the P.D across it will be Q/C and is opposite to the applied e.m.f. By Ohm's law, the effective e.m.f in the circuit will be IR = E − (Q/C). Applying the initial conditions i.e. Q = 0 when time t = 0, the above equation gives

$$Q = CE\,(1 - e^{-t/CR}) \text{ or } \frac{Q}{C} = E\,(1 - e^{-t/CR})$$

$$\therefore \qquad V = V_{max}\,(1 - e^{-t/CR})$$

where Q : charge on the capacitor at time t, V_{max}: applied e.m.f

After a time t = CR, the voltage becomes,

$$V = V_{max} (1 - e^{-t/CR}) = \frac{V_{max}}{e} = \left(1 - \frac{1}{2732}\right) V_{max} = 0.632 \times V_{max}$$

$$V = 0.632 \times V_{max} \qquad \qquad \text{... (1)}$$

Thus the RC time constant can be obtained as the time taken by the capacitor to charge to 63.2 % of its maximum value.

- **Discharging a capacitor :**

  When a capacitor of capacity C is charged to its peak potential V_{max}, and is allowed to discharge through a resistance R, after a time 't', the voltage across the capacitor becomes, $V = V_{max} e^{-t/CR}$

  After a time t = CR, the voltage becomes,

$$V = \frac{V_{max}}{e} = \frac{V_{max}}{2.738} = 0.368 \times V_{max}$$

$$V = 0.368 \times V_{max} \qquad \qquad \text{... (2)}$$

Thus the RC time constant can be obtained as the time taken by the capacitor to discharge to 36.8 % of its maximum value.

Precautions :

1. The voltmeter used should have high resistance.
2. Connect the circuit very carefully. Do NOT connect the terminals A and C. (This will short circuit the battery.)
3. Measure the theoretical value of R properly.
4. Start the stopwatch as soon as the keys are closed.
5. Take readings till the P.D. across the capacitor attains its maximum value (V_{max}).

Check Your Grasp :

1. *What do you mean by RC time constant?*
2. *What is the physical significance of RC time constant?*
3. *Define capacitance and state its units.*
4. *What are the factors on which the capacitance of a parallel-plate capacitor depends?*
5. *What is the phase relationship between voltage and current for a resistor?*
6. *What is the phase relationship between voltage and current for a capacitor?*

 **Calculations :**

STUDY OF LR CIRCUIT

Aim : To study LR circuit and to determine its characteristic parameters from voltage vector triangle.

Apparatus : A step down transformer with variable adjustment, Inductor, Capacitor, Resistor, A.C. voltmeter, AC milliammeter, Connecting wires, etc.

Figure :

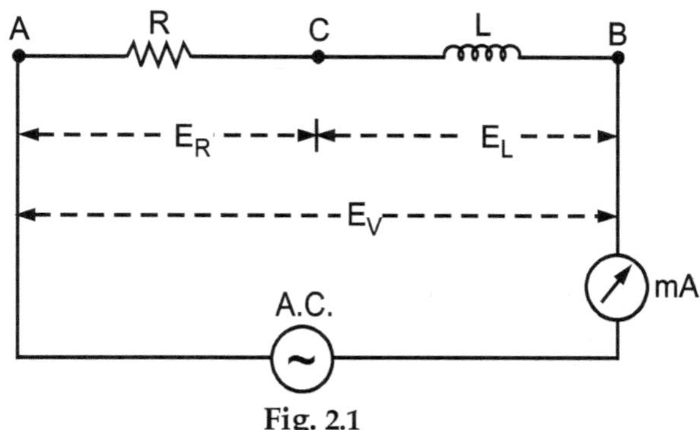

Fig. 2.1

Procedure :

1. Connect the circuit as shown in Fig. 2.1.
2. Connect the voltmeter across the points A and B, switch on the signal generator and adjust the applied voltage so that a suitable current flows through the circuit. Note the current I and voltage E_v.
3. Connect the voltmeter across R (between points A and C). Ensure constant current (I) through the circuit. Note the voltage across R i.e. E_R.
4. Connect the voltmeter across L (between points C and B). Ensure that the current through the circuit i.e. I remains constant. Note the voltage across L i.e. E_L.
5. Repeat steps (2) through (4) for three different values of current.
6. Plot the voltage vector triangle as discussed in theory section on a graph paper.

7. From the voltage triangle, calculate (i) Impedance (Z) of the circuit, (ii) Ohmic resistance (r) of the inductor, (iii) Inductive reactance (X_L) of the circuit, (iv) Inductance (L) of the inductor, (v) Phase angle (θ) by which voltage leads the current, (vi) Power factor ($\cos \theta$) of the circuit.

 **Observations :**

- Applied signal voltage, V = ... volt
- Applied signal frequency, f = ... Hz

 **Observation Table :**

Obs. No.	Current (I) (mA)	(A)	Voltage across AB (E_V) (volt)	Voltage across AC (E_R) (volt)	Voltage across CB (E_L) (volt)	Resistance $R = E_R / I$ (Ω)
1						
2						
3						

 **Formulae :**

With reference to voltage vector triangle as described in theory section :

(a) Impedance of the circuit,
$$Z = \sqrt{R^2 + X_L^2}$$

(b) Ohmic resistance (r) of the inductor,

(c) Inductive reactance (X_L), $X_L = \dfrac{CD}{AB} R$

(d) Inductance, $L = \dfrac{R}{2\pi f} \dfrac{CD}{AB}$

(e) Phase angle, θ : angle between E_V and E_R

(f) Power factor $= \cos \phi$

 **Results :**

- Impedance of the circuit   Z = ... Ω
- Ohmic resistance of the inductor   r = ... Ω
- Inductive reactance X_L = ... Ω
- Inductance of the inductor coil   L = ... mH
- Phase angle θ = ... degree
- Power factor of the circuit $\cos \theta$ = ...

Theory :

- When an alternating voltage is applied to a circuit containing an inductor (L) and resistor (R), the current flowing through the circuit is given by

$$I = \frac{E_v}{\sqrt{R^2 + X_L^2}} = \frac{E_v}{Z}$$

where, R : Resistance, $X_L = 2\pi f L$: inductive reactance at frequency f,

$Z = \sqrt{R^2 + X_L^2}$: impedance of the circuit

Thus impedance of the circuit and hence current through the circuit depends upon the applied frequency.

- In case of a purely resistive circuit, voltage and current are in phase with each other, in a purely capacitive circuit, voltage lags the current by $\pi/2$ radian and in case of purely inductive circuit voltage leads the current by $\pi/2$ radian.

- In an LR circuit, voltage leads the current by an angle θ, which can be determined by drawing voltage vector triangle.

- A practical inductor coil possesses some ohmic resistance and hence the net effect shown by it is the resultant of ohmic resistance and inductive reactance.

- Method to draw voltage vector triangle and obtain parameters from it:

(i) Using the measured values of E_V, E_R and E_L, draw a triangle as shown:

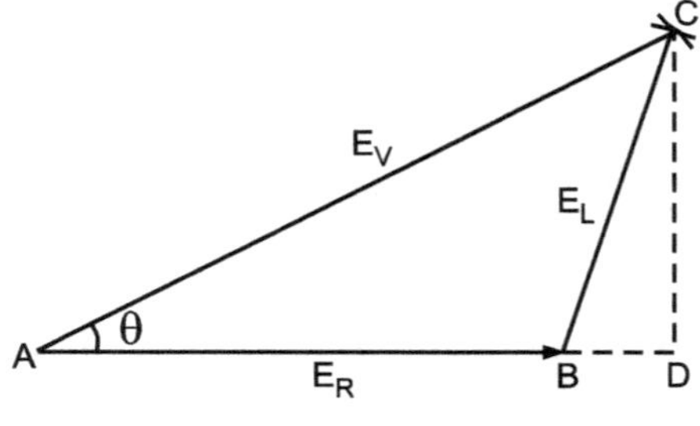

Fig. 2.2

(ii) Draw line AB; which represents E_R. With point A as centre and magnitude of E_V as radius, draw an arc. Similarly, with point B as centre and magnitude of E_L as radius, draw another arc. Mark the point of intersection as C. Join triangle ABC; which is the voltage vector triangle.

(iii) In $\triangle$ ABC, angle θ is the phase angle by which voltage leads the current.

(iv) Extend the lines from B and C to get triangle BCD. This triangle represents the voltage vector triangle for pure inductor; L. The segment BD represents voltage across ohmic resistance of the inductor coil, segment CD represents the voltage across the pure inductance and the resultant of these two give the net voltage across the inductor coil.

(v) Now, $E_R = AB = RI_v$ and $BD = rI_v$, where 'r' is the ohmic resistance of the inductance coil. Thus, $\dfrac{r}{R} = \dfrac{BD}{AB}$ or $r = \dfrac{BD}{AB} R$

(vi) Segment CD represent voltage across pure inductance. This can be obtained from the voltage triangle plotted on a graph paper. As $CD = X_L I_v = 2\pi f L I_v$.

$$\text{Thus} \qquad \frac{OD}{AB} = \frac{X_L}{R} = \frac{2\pi f L}{R} \qquad \text{or} \qquad X_L = \frac{CD}{AB} R$$

$$\text{and} \qquad L = \frac{R}{2\pi f} \qquad \text{or} \qquad L = \frac{CD}{AB}$$

- **Power factor:** Power factor is a measure of power losses in the reactive circuit. Higher is the value of power factor, more is the power loss in the circuit. It is given by $\cos\theta$.

Precautions :

1. Connecting wires should be properly insulated.
2. Do not touch any open (non-insulated) part of the circuit.
3. Choose suitable values of R and L.
4. Current through the circuit should remain constant throughout the experiment.

Check Your Grasp :

1. *What is a resistor?*
2. *What is an inductor?*
3. *What is ohmic resistance, inductive reactance and impedance of a circuit?*
4. *What is the phase relationship between voltage and currents in LR circuit?*
5. *What is an alternating current?*
6. *What is meant by r.m.s value of A.C?*
7. *What is meant by angle of lead and lag?*
8. *What is the effect of frequency of A.C on behavior of an inductor?*
9. *What is meant by voltage vector triangle?*
10. *What are the applications of LR circuit?*

Calculations :

STUDY OF LCR CIRCUIT

Aim : To study series LCR resonant circuit and to determine its characteristic parameters from resonance curve.

Apparatus : Signal generator, A.C. voltmeter, Resistors, Capacitor, Inductor, connecting wires, etc

Figure :

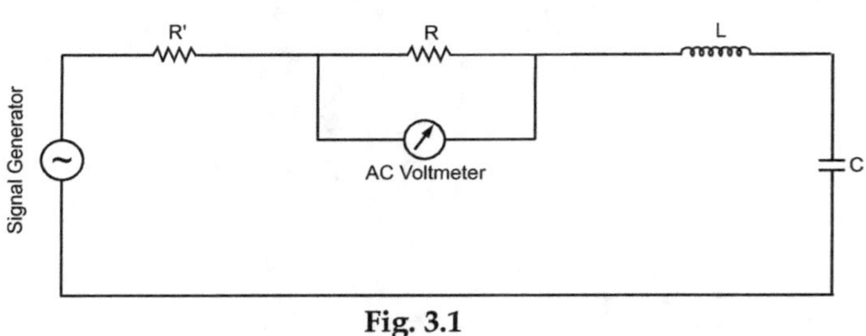

Fig. 3.1

Procedure :

1. Connect the circuit as shown in Fig. 3.1.

2. Adjust the output voltage of the signal generator to suitable value.

3. Change the signal frequency in suitable steps as indicated in the table and note down the voltage across the resistor **R.**

4. Calculate the current **I** flowing through the circuit.

5. Plot the graph of current **I** versus $\log_{10} f$.

6. Determine the resonant frequency of the series LCR circuit, theoretically and experimentally. Hence calculate the quality factor, impedance of the circuit at resonance and power factor of the circuit.

 **Observations :**

- $L =$ .... mH
- $C =$ .... μF

- $R =$ .... kΩ
- $R' =$ .... Ω

- $R_T = R + R'$ where R' is the output impedance of the signal generator.

 **Observation Table :**

Obs. No.	Frequency		Voltage (V) (volt)	Current (I) I = V/R (mA)	Obs. No.	Frequency		Voltage (V) (volt)	Current (I) I = V/R (mA)
	(f) (Hz)	log (f)				(f) (Hz)	log (f)		

 **Formulae :**

(a) Resonant frequency, $f_r = \dfrac{1}{2\pi\sqrt{LC}}$

(b) Quality factor (by graph), $Q = 2\pi \dfrac{f_r}{\Delta f}$ where $\Delta f = f_2 - f_1$: bandwidth

(c) Quality factor (by calculations), $Q = \dfrac{1}{R_T}\sqrt{\dfrac{L}{C}}$

(d) Impedance of the circuit, $Z = \sqrt{R^2 + (X_L - X_C)^2}$

(e) Power factor $= \cos\theta$

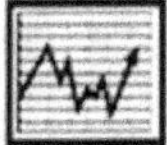

Graph :

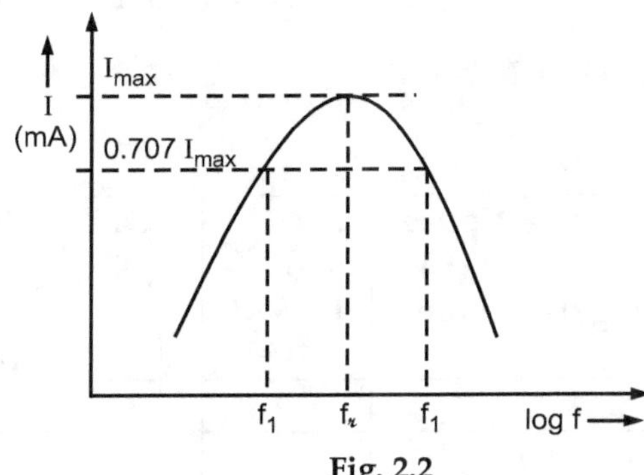

Fig. 2.2

Results :

- **Resonant frequency (by calculations)** f_r = ... Hz

- **Resonant Frequency (by graph)** f_r = ... Hz

- **Quality factor (by calculations)** Q = ...

- **Quality factor (by graph)** Q = ...

- **Impedance of the circuit at resonance** Z = ... Ω

- **Power factor of the circuit at resonance** $\cos\theta$ = ...

Theory :

- When an alternating voltage is applied to a circuit containing an inductor (L), capacitor (C) and resistor (R), the current flowing through the circuit is given by

$$I = \frac{E}{\sqrt{R^2 + (X_L - X_C)}} = \frac{E}{Z}$$

where, R : Resistance, $X_L = 2\pi fL$: inductive reactance at frequency f, $X_C = 1/2\pi fC$ capacitive reactance at frequency f : $Z = \sqrt{R^2 + (X_L - X_C)^2}$ impedance of the circuit, $\omega = 2\pi f$: angular frequency of applied A.C. signal. Thus impedance of

- the circuit and hence current through the circuit depend upon the applied frequency.
- In case of a purely resistive circuit, voltage and current are in phase with each other, in a purely capacitive circuit, voltage lags the current by $\pi/2$ radian and in case of purely inductive circuit, voltage leads the current by $\pi/2$ radian. In a series LCR circuit, the phase angle or phase difference, ϕ, between the voltage and current is given by, $\tan\phi = \dfrac{X_L - X_C}{R}$
- When the natural frequency of the circuit and the frequency of the applied signal are exactly same, i.e. $f_{natural} = f_{applied} = f_r$ (say), resonance takes place. f_r is known as resonant frequency.
- **At resonance :** $X_L = X_C$, therefore, $2\pi f_r L = \dfrac{1}{2\pi f_r C}$

(i) **Resonant frequency:** $f_r = \dfrac{1}{2\pi\sqrt{LC}}$

(ii) **Impedance:**

Impedance of the circuit is given by $Z = \sqrt{R^2 + (X_L - X_C)^2}$

At resonance, $X_L = X_C$, hence $Z = R$ only i.e. *at resonance*, $I = \dfrac{E}{Z} = \dfrac{E}{R}$

Thus, at resonance, **impedance** of the series resonant circuit is **minimum** and hence *current* is **maximum.**

(iii) **Phase angle:**

Using $X_L = X_C$, $\tan\phi = \dfrac{X_L - X_C}{R} = \dfrac{0}{R} = 0$

Thus, at resonance, the current and voltage in a series resonant circuit have zero phase difference between them i.e. they are in phase with each other.

(iv) **Power factor:**
- Power factor of a circuit is given by $\cos\theta$. In a series LCR circuit, as $X_L = X_C$ (at resonance), power factor is unity (i.e. 1).
- Power factor is a measure of power losses in the reactive circuit. Higher is the value of power factor, more is the power loss in the circuit.

(v) **Quality factor:**

The ratio of potential drop across the capacitor or inductor to the P.D. across the resistor at resonance is called quality factor, Q.

For inductor, $Q_L = \dfrac{E_t}{E_R} = \dfrac{\omega L}{R}$ and for capacitor, $Q_L = \dfrac{E_C}{E_R} = \dfrac{1}{\omega CR}$

At resonance, $X_L = X_C$, hence $Q_L = Q_C = Q$

Also, $\omega L = \dfrac{1}{\omega C} \Rightarrow \omega = \dfrac{1}{\sqrt{LC}}$

$$\therefore \quad Q = \frac{\omega L}{R_T} = \frac{\frac{1}{\sqrt{LC}} \times L}{R_T} = \frac{1}{R_T} \times \frac{L}{\sqrt{LC}} = \frac{1}{R_T} \times \sqrt{\frac{L^2}{LC}} = \frac{1}{R_T} \times \sqrt{\frac{L}{C}}$$

$$\therefore \quad Q = \frac{1}{R_T} \times \sqrt{\frac{L}{C}}$$

From the graph of current versus frequency, quality factor Q can be calculated as

$Q = \dfrac{f_r}{\Delta f}$ where, $\Delta f = f_2 - f_1$ is the bandwidth and f_r is the resonant frequency.

 **Precautions :**

1. The values of inductance L and capacitance C should be selected in such a way that the natural frequency of the series LCR circuit is around 10 kHz.

2. The value of output voltage of the oscillator and the resistance must be kept constant throughout the experiment.

3. Measure the frequency and bandwidth very carefully.

 **Check Your Grasp :**

1. *What is resonance?*

2. *State types of resonant circuits and explain the basic difference between them.*

3. *What do you mean by inductive reactance, capacitive reactance and impedance of a circuit?*

4. *What is the phase relationship between voltage and currents in case of a resistor, a capacitor and an inductor?*

5. *Define the terms: (i) Resonant frequency, (ii) Phase, (iii) Power factor, (iv) Quality factor*

6. *What is meant by series resonance? How it differs from parallel resonance?*

7. *What is meant by sharpness of resonance?*

8. *Define the term: Bandwidth.*

9. *What is the significance of power factor?*

10. *What are the applications of a series resonant circuit?*

 **Calculations :**

❖ ❖ ❖

STUDY OF KIRCHHOFF'S LAWS

 **Aim :**   To verify Kirchhoff's current and voltage laws.

 **Apparatus :**   Power supply, Voltmeter, Milliammeter, Resistors, Connecting wires etc.

 **Figure :**

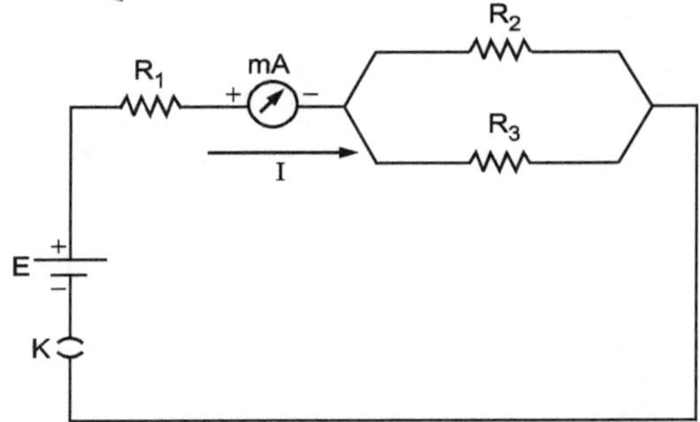

Fig. 4.1 : To measure current entering the junction (I)

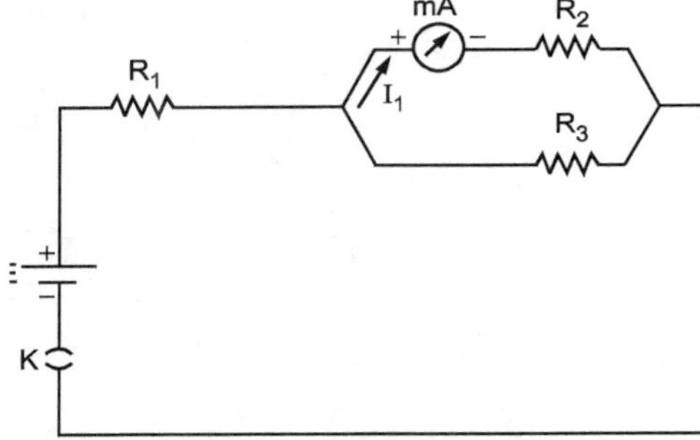

Fig. 4.2 : To measure current leaving the junction (I_1)

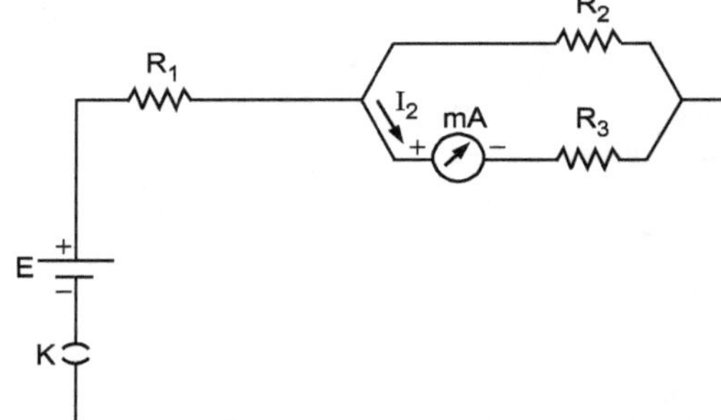

Fig. 4.3 : To measure current leaving the junction (I_2)

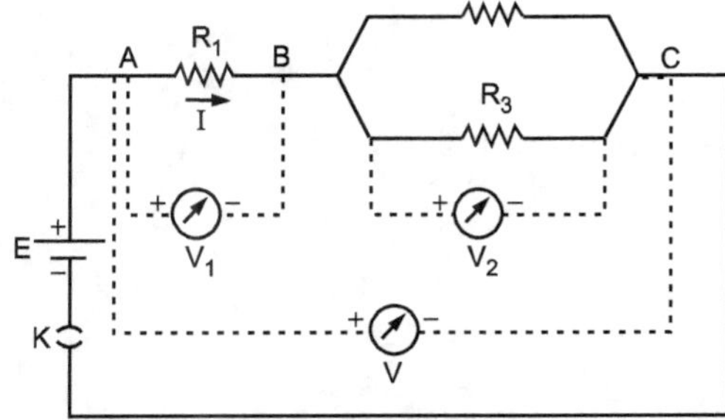

Fig. 4.4 : To verify Kirchhoff's voltage law

Procedure :

Part I (Kirchhoff's Current Law)

1. Connect the circuit as shown in Fig. 4.1.
2. Connect the milliammeter as shown in Fig. 4.1 and note the current entering the junction i.e. I, for the applied voltages 3 V, 6 V and 9 V successively.
3. Connect the milliammeter as shown in Fig. 4.2 and note the current leaving the junction i.e. I_1, for the applied voltages 3 V, 6 V and 9 V successively.
4. Connect the milliammeter as shown in Fig. 4.3 and note the current leaving the junction i.e. I_2, for the applied voltages 3 V, 6 V and 9 V successively.

Part II (Kirchhoff's Voltage Law)

1. Connect the circuit as shown in Fig. 4.4.
2. Note the voltage across the points A and C (V), points A and B (V_1) and across the points B and C (V_2) for the applied voltages 3 V, 6 V and 9 V successively; by connecting the voltmeter at appropriate positions.

Observation Table :

Part I : To verify Kirchhoff's Current Law (KCL) :

Obs. No.	Applied Voltage (volt)	Current entering the junction (I) (mA)	Current leaving the junction (I_1) (mA)	Current leaving the junction (I_2) (mA)	$I' = I_1 + I_2$ (mA)
1	3 V				
2	6 V				
3	9 V				

If I = I', then Kirchhoff's Current Law is verified.

Part II : To verify Kirchhoff's Voltage Law (KVL) :

Obs. No.	Applied Voltage (volt)	Voltage across points A and C (V) (volt)	Voltage across points A and B (V_1) (volt)	Voltage across points B and C (V_2) (mA)	$V' = V_1 + V_2$ (volt)
1	3 V				
2	6 V				
3	9 V				

If V = V', then Kirchhoff's Voltage Law is verified.

Conclusions :

Kirchhoff's Current and Voltage laws are experimentally verified.

Theory :

- **Methods to connect the components (series - parallel) :** Electronic components can be connected in series or parallel combination.
- When current has only one path to travel through the components, those components are connected in series.
- When current has more than one path to flow, those components are connected in parallel.
- In case of simple electrical circuits containing a battery and resistances in series and parallel, the current and potential difference across any resistance is obtained by using Ohm's law. In case of complicated circuits, the calculations are done using Kirchhoff's laws.
- There are two laws given by Kirchhoff; (i) Kirchhoff's Current Law (KCL) and (ii) Kirchhoff's Voltage Law (KVL).
- **Kirchhoff's Current Law (KCL) :** In any network of conductors in an electrical circuit, the algebric sum of currents meeting at any point (junction) is zero.
  i.e. $\Sigma I = 0$
  In Fig. 4.5, I_1 and I_5 are the currents entering the junction and I_2, I_3, I_4 are the currents leaving the junction. Then according to KCL,

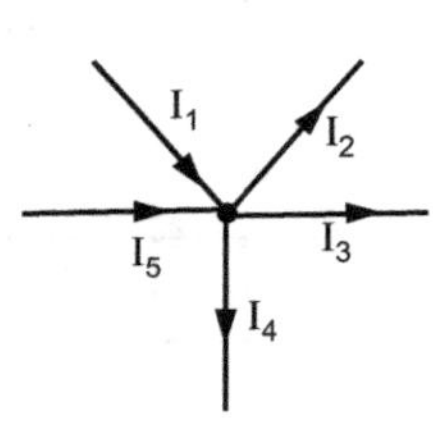

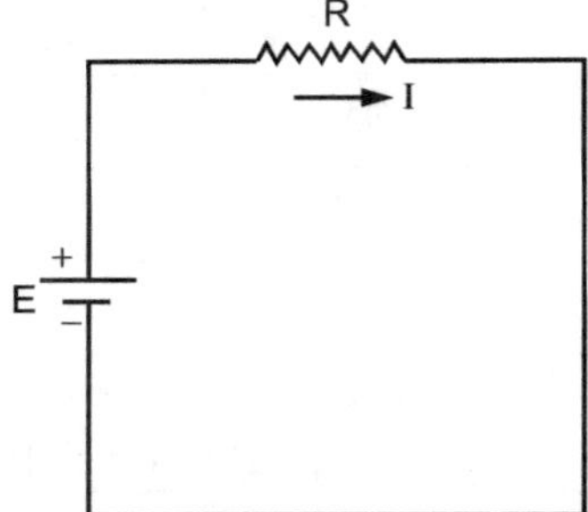

Fig. 4.5 : Kirchhoff's Current Law   Fig. 4.6 : Kirchhoff's Voltage Law

$$I_1 + I_5 - I_2 - I_3 - I_4 = 0$$

i.e. $$I_1 + I_5 = I_2 + I_3 + I_4$$

The currents entering the junction are taken as positive whereas the currents leaving the junction are taken as negative.

- **Kirchhoff's Voltage Law (KVL) :** In a closed path of network of conductors, the algebraic sum of the products of resistance and resistance of each part of the closed path is equal to the algebraic sum of e.m.f's in the circuit.

- For applying KVL, the circuit is 'traced' round any closed path of conductor. The algebraic sum of the potential drop (IR) across the resistors is equal to the algebraic sum of the e.m.f's in that closed loop (Fig. 4.6).

i.e. $$\Sigma IR = \Sigma E$$

If the e.m.f's and currents are taken in the anticlockwise direction, they are taken as positive, those in the clockwise direction are taken as negative.

About the experiment:

- To verify KCL, a network is chosen and the current entering and leaving the junction are measured using current meter. When sum of incoming current(s) come out to be equal to sum of current(s) leaving that junction, KCL is verified.
- To verify KVL, a network is chosen and the potential drop across each resistor and the applied e.m.f are measured using voltmeter. When sum of applied e.m.f. (s) come out to be equal to sum of the product of current(s) and resistances, then KVL is verified.

 **Precautions :**

1. Keep key (K) open while changing the applied voltage.
2. Milliammeter should always be connected in series, whereas voltmeter should always be connected in parallel.

 **Check Your Grasp :**

1. *Why a current meter is connected in series with a network component to measure current through?*
2. *Why a voltmeter is connected in parallel across a network component to measure the potential drop across it?*
3. *State Kirchhoff's current and voltage laws.*
4. *State the sign conventions in case of KCL and KVL.*
5. *What are the advantages of Kirchhoff's law over Ohm's law?*
6. *What is meant by a junction?*
7. *What will be the effect if the polarities of the battery in the circuit are reversed?*
8. *State the applications of Kirchhoff's laws.*

Calculations :

DIODE CHARACTERISTICS

Aim : To determine the forward and reverse biased characteristics of a semiconductor diode.

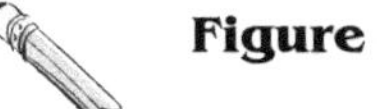

Apparatus : Semiconductor diode (1N4001 or BY127), Power supply, Current meter, Voltmeter, Rheostat, Resistance, Connecting wires etc

Figure :

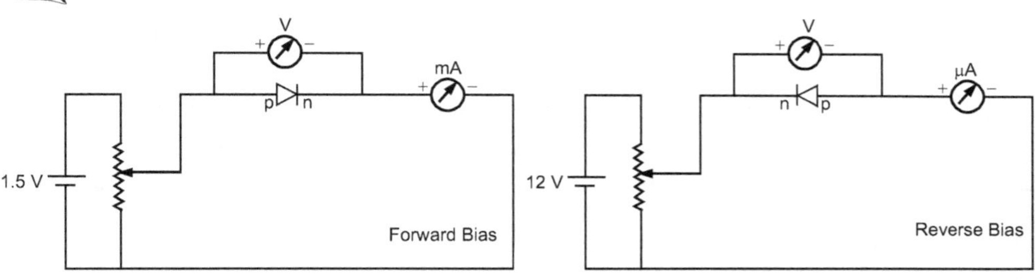

Fig. 5.1   Fig. 5.2

Procedure :

(A) Forward bias characteristics:

1. Connect the circuit as shown in Fig. 5.1.
2. Adjust the rheostat such that voltmeter reads zero.
3. Increase the voltage in steps as indicated in the observation table and note the corresponding milliammeter reading.
4. Plot the characteristic plot of I versus V and from the slope of the plot, calculate the forward bias resistance of the diode.

(B) Reverse characteristics:

1. Connect the circuit as shown in Fig. 5.2.
2. Adjust the rheostat such that the voltmeter reads zero.

3. Increase the voltage in steps of 1.0 volt and note the corresponding micro-ammeter reading.

4. Plot the characteristic plot of I versus V and from the slope of the plot, calculate the reverse bias resistance of the diode.

 **Observation Table :**

Obs. No.	Forward Bias			Reverse Bias		
	Voltage (V) (volt)	Current (I) (mA)	Current (I) (A)	Voltage (V) (volt)	Current (I) (μA)	Current (I) (A)
1						
2						
3						
4						
5						
6						
7						
8						
9						
10						

 **Formulae :**

(a) Diode resistance (Forward bias), R_d = (1/slope) of forward characteristics.

(b) Diode resistance (Reverse bias), Rr = (1/slope) of reverse characteristics.

 **Graph :**

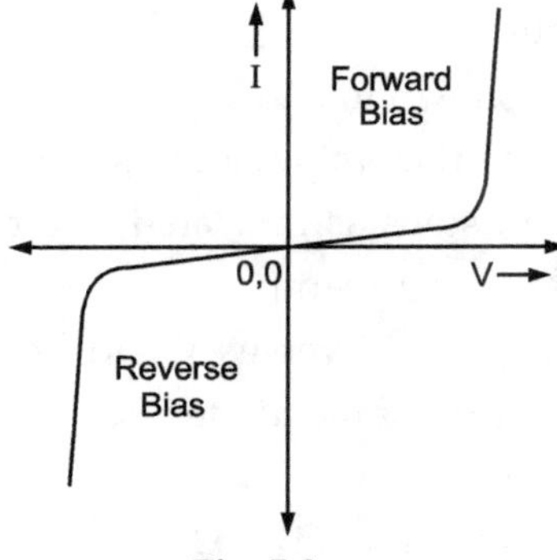

Fig. 5.3

Results :

- **Forward and Reverse bias characteristic plots of the diode are plotted.**
- **Forward bias diode resistance = ... Ω**
- **Reverse bias diode resistance = ... Ω**

Theory :

- According to band theory of solids, materials are classified as conductors, semiconductors and insulators. Energy gap or the forbidden energy gap between the conduction band and the valence band is minimum in case of conductors, maximum for insulators whereas moderate in semiconductors.

- **Types of semiconductors:** Semiconductors are of two types; extrinsic and intrinsic. The extrinsic semiconductors are of p and n type. In p-type semiconductors, holes are the majority current carriers whereas in n-type, electrons are the majority current carriers.

- **Diode:** When p and n type blocks of a semiconductor are joined together, it forms a diode.

- **Biasing a diode:** A diode is to be biased for its proper working. A diode has two types of biasing. When p block is connected to positive terminal (and n connected to negative terminal) of a battery, the diode is said to be forward biased and when p block is connected to negative terminal (and n connected to positive terminal), the diode is reverse biased.

- A diode has very low resistance (of the order of 100 Ω) in forward bias, hence it conducts heavily. Reverse biased diode has a very large resistance (of the order of 10^5 Ω) and normally does not conduct in reverse bias.

- The forward potential at which a diode suddenly conducts heavily is called its barrier potential or knee voltage.

- The reverse voltage at which the diode conducts due to breakdown mechanism is called reverse breakdown voltage.

- The forward and reverse bias characteristic plots are as shown in Fig. 5.3.

Precautions :

1. Connect the circuit as given in Fig. 5.3 and then switch ON the power supply.
2. Initially, adjust the position of rheostat so as to read zero voltage across the diode.
3. Switch OFF the power supply immediately on the reverse breakdown of the diode.
4. Do not make any changes in circuit connections till the power supply is ON.

Check Your Grasp :

1. *How solids are classified on the basis of band theory?*
2. *What do you mean by a semiconductor? What are its types? Give their examples.*
3. *What do you mean by p and n type semiconductors? How they are formed?*
4. *Explain the concept, doping, majority & minority carrier, potential barrier, breakdown voltage, depletion layer and biasing of a diode.*
5. *What is a semiconductor diode? What are its applications?*
6. *With proper reasoning, explain the forward and reverse bias characteristics of a semiconductor diode.*
7. *What do you mean by knee voltage and breakdown voltage?*
8. *What is meant by forbidden energy gap?*
9. *Explain the meaning of static and dynamic resistances with reference to a semiconductor diode.*
10. *State any three applications of a semiconductor diode.*

Calculations :

STUDY OF VOLTMETER, AMMETER & MULTIMETER (AC/DC RANGES & LEAST COUNT)

Aim :   To study and measure voltage-currents (ac/dc) using multimeter, voltmeter and ammeter.

Apparatus :   Multimeter (analogue/digital), 1.5 V dry cell, Resistances, Variable ac/dc source, Connecting wires, Chords etc.

Figure :

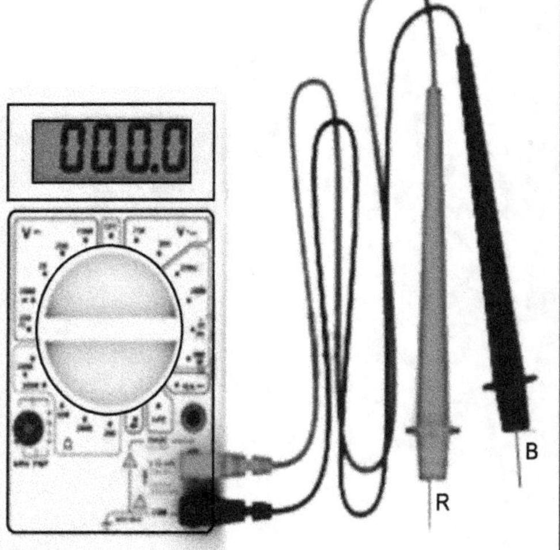

Fig. 6.1 : Digital Multimeter

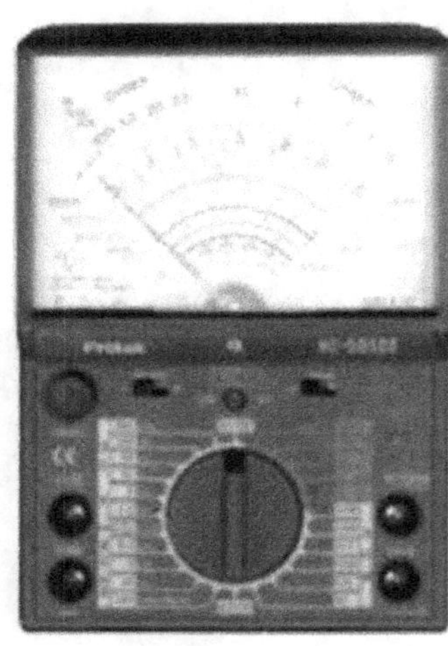

Fig. 6.2 : Analogue Multimeter

Procedure :

Measurement of AC/DC voltages:

1. Connect the output of the DC battery/Dimerstat (AC) to the multimeter/voltmeter (DC/AC) and switch on battery/AC mains, multimeter and ac/dc voltmeter.
2. Note down the various ranges given on DMM or voltmeter also the smallest measurement in that prescribed range.
3. In case of dc voltage measurement vary the battery voltage and measure the corresponding dc voltage on DMM and DC voltmeter by changing various ranges given on it.
4. In case of ac voltage, vary the dimerstat voltage and measure the corresponding output voltage on multimeter, AC voltmeter by changing various ranges given on multimeter and AC voltmeter.

Measurement of AC/DC currents:

1. Connect the output of the DC battery DMM/ DC current meter through 1 KΩ resistance in series. Make similar adjustment for the AC current measurement.
2. Switch on DMM, AC/DC current meter AC mains/battery.
3. Note down the various ranges given on DMM or ammeter also the smallest measurement in that prescribed range.
4. Vary input given to the DMM or AC/DC ammeter and measure the corresponding AC/DC current for various ranges given on it.

Observation Table :

Table (1) Measurement of AC/DC voltage

Obs. No.	Range selected	Resolution/least count	Observed voltage (volt)	
			DMM	Voltmeter (DC/AC)
DC voltage measurement				
1.	0 - 10 V DC			
2.	0 - 50 V DC			
3.	0 - 100 V DC			
AC voltage measurement				
4.	0 - 10 V AC			
5.	0 - 100 V AC			
6.	0 - 1000 V AC			

Table (2) Measurement of AC/DC current

Obs. No.	Range selected (AC/DC)	Resolution/ Least count	Observed AC/DC currents (mA)	
			DMM	Ammeter
DC Current measurement				
1.	0 – 50 µA			
2.	0 – 10 mA			
3.	0 – 500 mA			
AC Current measurement				
1.	0 – 50 µA			
2.	0 – 10 mA			
3.	0 – 500 mA			

 **Results :**

- DC/AC voltages and current were measured using DMM and voltmeter/ ammeter.

- The measured values using DMM and voltmeter/ammeter are consistent with each other.

 **Theory :**

- Voltmeter is an electronic measuring instrument used for measuring electrical potential difference between two points in an electric circuit.

- **Principle of voltmeter:** It is always connected in parallel to the component where the voltage is to be measured. Parallel connection is used because a voltmeter is constructed in such a way that it has a very high value of resistance. If it is connected series, due to high resistance, zero current would flow through it which means the circuit becomes open.

- Ammeter is an electronic instrument used for the measurement of current passing through a component.

- **Principle of ammeter:** It is always connected in series with the component where the current passing through it is to be measured. Series connection is used because ammeter is constructed in such a way that it has a very low value of resistance.

- Multimeter is an electrical device which directly measures or indicates various electrical parameters such as voltages (ac/dc), current (ac/dc), resistances, continuity in the circuit, capacitances (specially designed), etc.

- It consists of moving coil galvanometer (MCG) which is sensitive to lower values of currents, voltages and resistances and having various scale measurements shown on its front panel.

- To use it as a current measurement device, fixed values of resistances are connected in parallel to the galvanometer coil called as shunt resistance which can measure various range of currents in the circuit.

- To use it as a voltage measuring device, few more resistances of fixed values are provided which are connected in series with the galvanometer to measure the desired potential difference in the circuit.

- A dry cell of around 1.5 V is used in the multimeter for measuring resistance.

- The input is provided to the multimeter by using two lead pins i.e. one is red for positive terminal and the other is black for ground purpose.

- Generally few switches are provided on the multimeter which vary in numbers w.r.t. different designers and models. Few of the common are range switch, function switch and zero ohm switches.

- Also various holes are provided on the front panel normally below the scale arrangement. These are also called as sockets which are used for different types of measurements such as voltages, currents, etc.

- *Voltage* is the measure of electrical work done to move electrons through a conductor. It is measured in terms of volts.

- *Current* is the rate of flow of charges and it is measured in terms of ampere.

 **Precautions :**

1. For measuring ac/dc voltages and currents, set the respective function switch. If the selected switch is reversed then there is a possibility of damage of the multimeter.

2. Always connect voltmeter across the component and current meter (Ammeter) in series.

3. For accurate measurement of current/voltage/resistance, select the switch at the proper range.

4. When voltage/current is measured, do not touch the leads.

5. Always connect the black lead to ground/common or negative terminal.

Check Your Grasp :

1. *What is meant by voltage and current?*

2. *Why voltmeter connected across the component?*

3. *What is meant by a multimeter?*

4. *What are the different types of multimeters?*

5. *What are the various parameters that can be measured with a multimeter?*

6. *Is it possible to measure capacitance and inductance using a multimeter?*

7. *Is it necessary to calculate least count of a multimeter?*

8. *Why there is a difference in theoretical and measured value of resistance?*

9. *What is the significance of tolerance in the resistance of resistor?*

10. *What is the difference between galvanometer and voltmeter?*

11. *What is the difference between galvanometer and current meter?*

12. *How galvanometer can be used as a voltmeter and ammeter in a circuit?*

13. *What it means when voltmeter and ammeter shows negative deflections?*

14. *How analogue ac and dc voltmeters/ammeters are identified?*

15. *If only ammeter is connected in a circuit then is it possible to calculate voltage?*

Calculations :

DETERMINATION OF FREQUENCY OF AC MAINS

Aim : To determine the frequency of AC mains using sonometer and vibrating wire.

Apparatus : Sonometer, Step down transformer, Slotted weights, Two strong bar magnets, Rheostat, Connecting wires etc.

Figure : To determine the frequency of A.C. mains using vibrating wire of a sonometer.

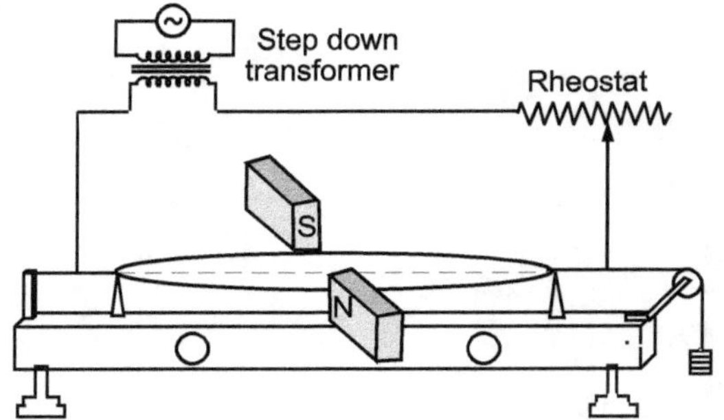

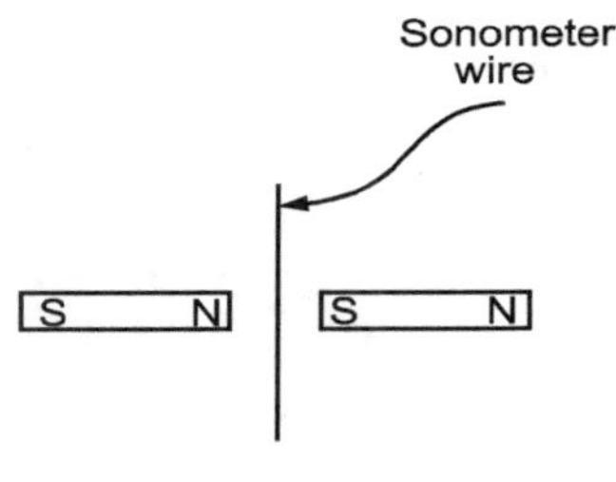

Fig. 7.1 : Sonometer arrangement   Fig. 7.2 : Magnet arrangement

Procedure :

1. Make the arrangement and connect the circuit as shown in Fig. 7.1.
2. Add a mass of 50 g to the hanger. Switch ON the power supply.
3. Adjust length of the sonometer wire using knife edges so that the wire will vibrate vigorously.
4. Note the resonating length of the wire (distance between the two knife edges).
5. Repeat steps (3) and (4) for masses 100 g, 150 g, 200 g and 250 g. Measure the resonating length of the sonometer wire for each mass.
6. Measure the radius of the sonometer wire using micrometer screw gauge.
7. Plot the graph of ℓ versus $\sqrt{T}$ and carry out the calculations as indicated.

 **Observations :**

- Least count of the micrometer screw gauge = ... cm
- Diameter of the wire

  (a) d_1 = ... cm

  (b) d_2 = ... cm

  (c) d_3 = ... cm

  Mean diameter, d = $\dfrac{d_1 + d_2 + d_3}{3}$ = ... cm

  Radius of the wire, r = ... cm

- Density of material of the wire, ρ = ... g/cc
- Mass per unit length of the wire, m = $\pi r^2 \rho$ = ... g/cm

 **Observation Table :**

Obs. No.	Mass attached (M) g	Tension T = Mg dyne	$\sqrt{T}$	Resonating length (ℓ)				$\sqrt{T}/\ell$	Mean $\sqrt{T}/\ell$
				Trial I cm	Trial II cm	Trial III cm	Mean cm		
1	50								
2	100								
3	150								
4	200								
5	250								
6	300								

 **Formulae :**

Frequency of A.C. mains,

$$N = \frac{1}{2\ell}\sqrt{\frac{T}{m}} \Rightarrow N = \frac{1}{2\sqrt{m}}\left(\frac{\sqrt{T}}{\ell}\right)_{mean} \Rightarrow N = \frac{1}{2\sqrt{m}}\left(\frac{1}{slope}\right) Hz$$

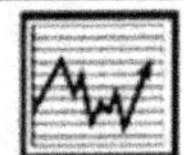

Graph :

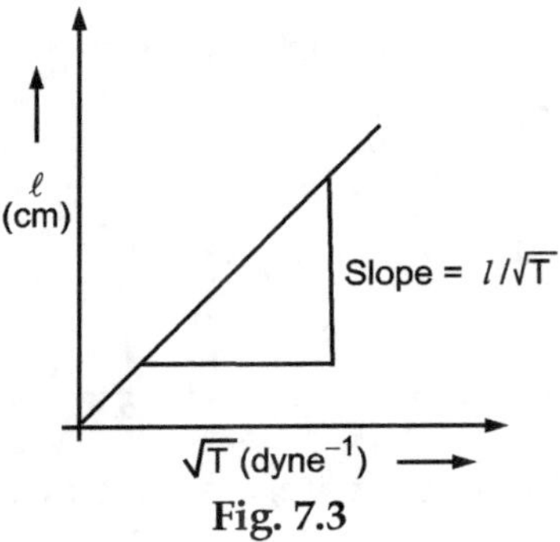

Fig. 7.3

Result :

- Frequency of A.C. mains (by calculations), N = ... Hz
- Frequency of A.C. mains (by graph),   N = ... Hz

Theory :

- **Frequency:** The number of cycles described by a wave per unit time is called frequency. Its S.I unit is Hz.
- **Direct current (dc):** The current provided by a battery has a continuous flow only in one direction with constant magnitude at all the time. This is called direct current or d.c.
- **Alternating current :**
  - The current from the mains supply is sinusoidal in nature; its magnitude varies with time. This is called alternating current or A.C.
  - A.C. is represented by $I = I \sin \omega t$, where ω is the angular frequency, with $\omega = 2\pi f$, f being the frequency of A.C.

About the experiment:

- When A.C. is passed through the sonometer wire, according to Oersted's experiment, alternating magnetic field is developed about the wire.
- When two bar magnets are held horizontally in the plane of the wire, the wire is alternately repelled away from the magnet. These vibrations produce a stationary wave pattern along the length of the wire, just like in case of Melde's experiment. When the natural frequency of vibration of the sonometer wire is exactly same as that of A.C. mains frequency, it vibrates vigorously with the maximum amplitude.

- When one single loop is formed, the wire is said to be vibrating in the fundamental mode of vibration with nodes at the two knife edges and antinode at the centre. According to the theory of vibrating string, frequency of vibrations of the wire is given by

$$N = \frac{n}{2l}\sqrt{\frac{T}{m}} \text{ Hz}$$

  where ℓ is the length of the loop, n is the number of loops, m is the mass per unit length of the wire and T is the tension in the wire.

- In fundamental mode of vibration, the number of loops become one i.e. n = 1.

Precautions :

1. Sonometer wire should be straight (without any kinks) and of uniform thickness to have uniform mass per unit length.
2. Adjust the length of vibrating wire in such a way that the loop becomes well defined.
3. The pulley should be frictionless.
4. Measure the resonating length accurately.

Check Your Grasp :

1. *What is an alternating current?*
2. *Distinguish between A.C and D.C*
3. *What do you mean by A.C. mains? What is its frequency in India?*
4. *What is meant by resonance?*
5. *How resonance takes place in this experiment?*
6. *Why loop is formed along the length of vibrating wire?*
7. *If the number of loops along the stretched string is more than one, how will you determine the frequency?*
8. *What is the relation between mass attached and number of loops formed?*
9. *Why the wire vibrates in the vertical plane?*
10. *State the laws of vibrating string.*
11. *State standard AC mains voltage and frequency in India and in USA.*

 **Calculations :**

COMPARISON OF CAPACITOR USING DESAUTY'S METHOD

 **Aim :** To compare the capacities of the given condensers by Desauty's method and determine the value of unknown capacitance.

 **Apparatus:** Two condensers, Two resistance boxes, Charge-discharge plug key, Battery, Spot galvanometer, connecting wires etc.

 **Figure :**

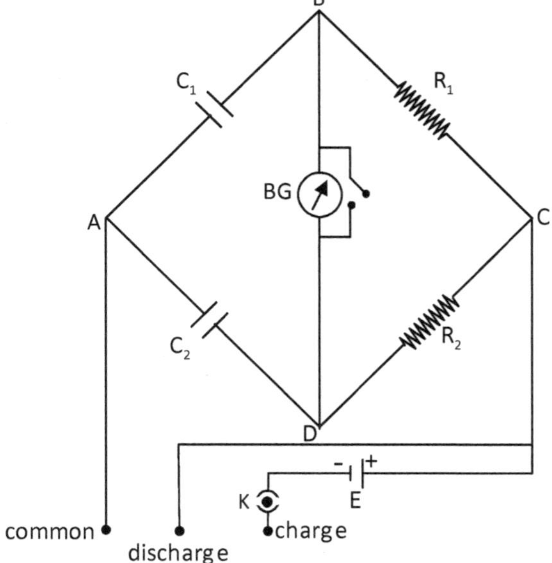

Fig. 8.1

Procedure :

- Connect the circuit as shown in Fig. 8.1.
- The resistance 1000 ohm is included in R_1 and 10,000 ohms in R_2. After charging and discharging with the key, throw on the scale is observed. Next, the resistances are interchanged. Again the key is pressed and released as before. The throw should be on the opposite side now.
- Set resistance R_1 as 5000 Ω (say) and adjust the resistance in R_2 such that the BG spot remains at the same position; even after charging and discharging. After every trial, the condensers should be discharged with a piece of wire so that for the next trial, the spot would not be disturbed during the charging. Note down the value of R_2.
- Repeat the experiment for different values of R_1 (say 6000, 7000 Ω etc.) and note down corresponding values of R_2 for no deflection.
- Using the formula, determine the value of unknown capacitance, C_2.

Observation Table :

$C_1 = \ldots\ldots \; \mu F$

Sr. No.	R_1 (Ω)	R_2 (Ω)	$C_2 = C_1\left(\dfrac{R_1}{R_2}\right)$
1			
2			
3			
4			
		Mean $C_2 =$	$\ldots\ldots\ldots \; \mu F$

Formulae :

$$\frac{C_1}{C_2} = \frac{R_2}{R_1}$$

where C_1 and C_2 are the capacities of the condensers and R_1 and R_2 are the resistances unplugged from the resistance boxes.

Result :

Capacities of the given condensers are compared using Desauty's method and the value of unknown capacitance $C_2 = $ .... μF.

Theory :

- **Condenser (Capacitor)**: It consists of two metal plates placed parallel to each other, space between these two plates being filled with air or some dielectric material (air, plastic, mica etc.)
- Adding electrical energy to a capacitor is called charging; releasing the energy from a capacitor is known as discharging. Capacitor is used to store the electrical energy.
- The capacity of capacitor is calculated by formula: $C = q/V$
  where q: charge on each plate and V is the potential difference between two plates.
- The dielectric constant (ε) of a dielectric material is the ratio of the capacity of a condenser with a dielectric material (C) to the capacity of the same condenser without the dielectric (C_0): $\varepsilon = \dfrac{C}{C_0}$
- Unit of capacitance is Farad (F) and $1\mu F = 10^{-6}$ F, $1pF = 10^{-12}F$

Desauty's Bridge Circuit:
- Desauty's Bridge is balanced (BG shows null deflection) when the condition, $C_1R_1 = C_2R_2$ is satisfied. Hence, ratio of capacitances of two condensers can be expressed as: $\dfrac{C_1}{C_2} = \dfrac{R_2}{R_1}$
- Thus, charges on the two condensers will grow at the same rate only if the time constants of the two circuits are equal.

Precautions :

1. The values of R_1 and R_2 should be high.
2. For sufficient sensitiveness of the bridge, the battery should have high E.M.F.
3. The connecting wire should not be connected loose and care should be taken while handling the condensers.

Check Your Grasp :

1. *What do you mean by condenser and when condenser is discharged?*
2. *What do you mean by capacity of a conductor?*
3. *What is the effect of dielectric constant on the capacity of a condenser?*

4. *What do you mean by dielectric material?*
5. *Does this method give accurate result?*

 **Calculations :**

www.ingramcontent.com/pod-product-compliance
Lightning Source LLC
LaVergne TN
LVHW080520200726
843508LV00005B/1433